AF539276

Contents

11 **Foreword**
By Poul Erik Tøjner

20 **The Moon**
– From Inner Worlds to Outer Space
By Marie Laurberg

40 **Lunar Influence**
By E.C. Krupp

49 **Moon Poems**
By Jorge Luis Borges, Sappho, Emily Dickinson, Li Po, Percy Bysshe Shelley, Sylvia Plath, Walt Whitman, D.H. Lawrence and Fernando Pessoa

58 **The Distance of the Moon**
By Italo Calvino

84 **Astronautic Theater:**
Space Flights and Lunar Expeditions in 1960s Art
By Stephen Petersen

108 **The Scientific Moon**
By Anja C. Andersen

121 **List of Works**

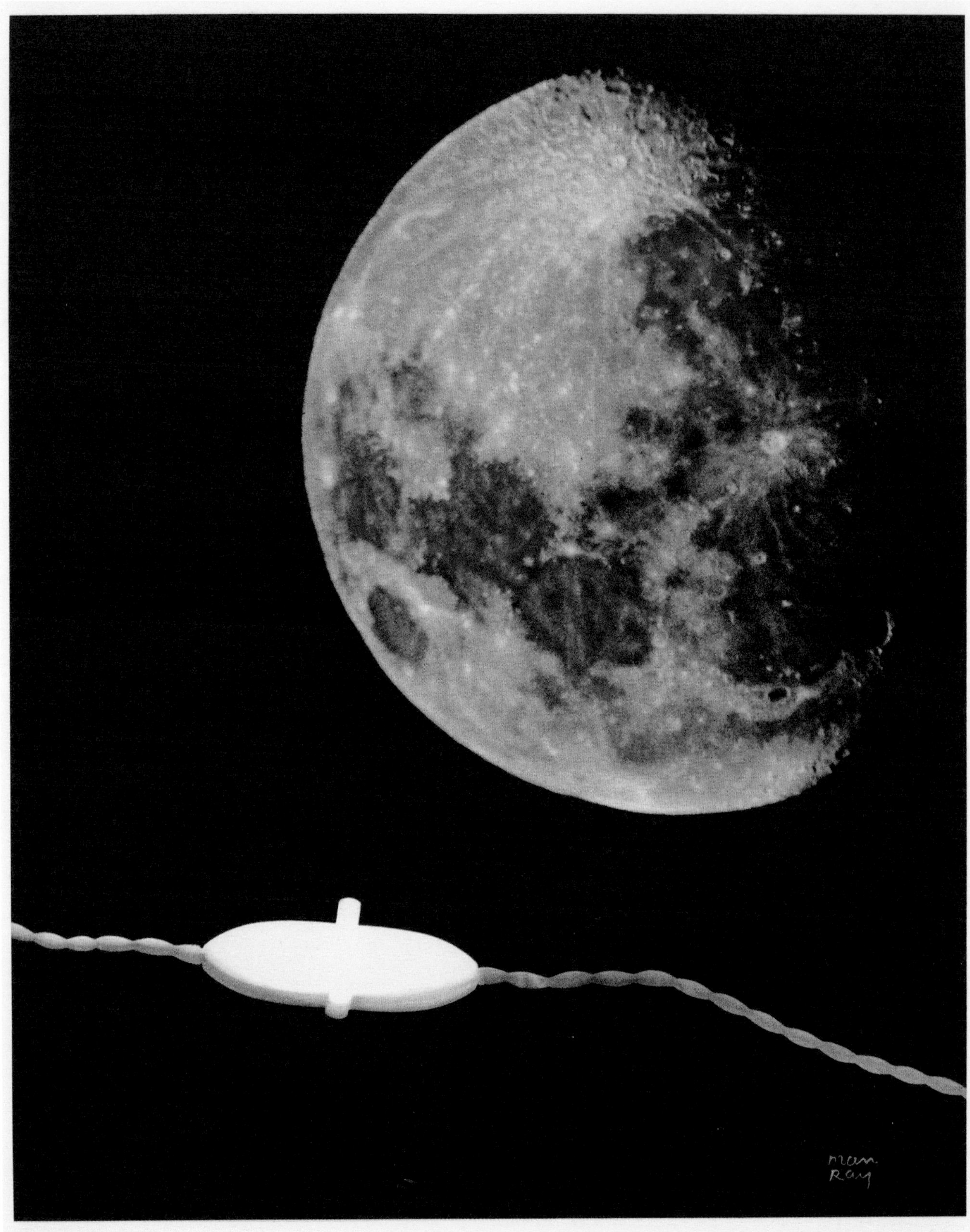

cat. 139
Man Ray
Le Monde, 1931
The World

Astronomer
Johann Friedrich
Julius Schmidt's
plaster model of the
moon, 1898

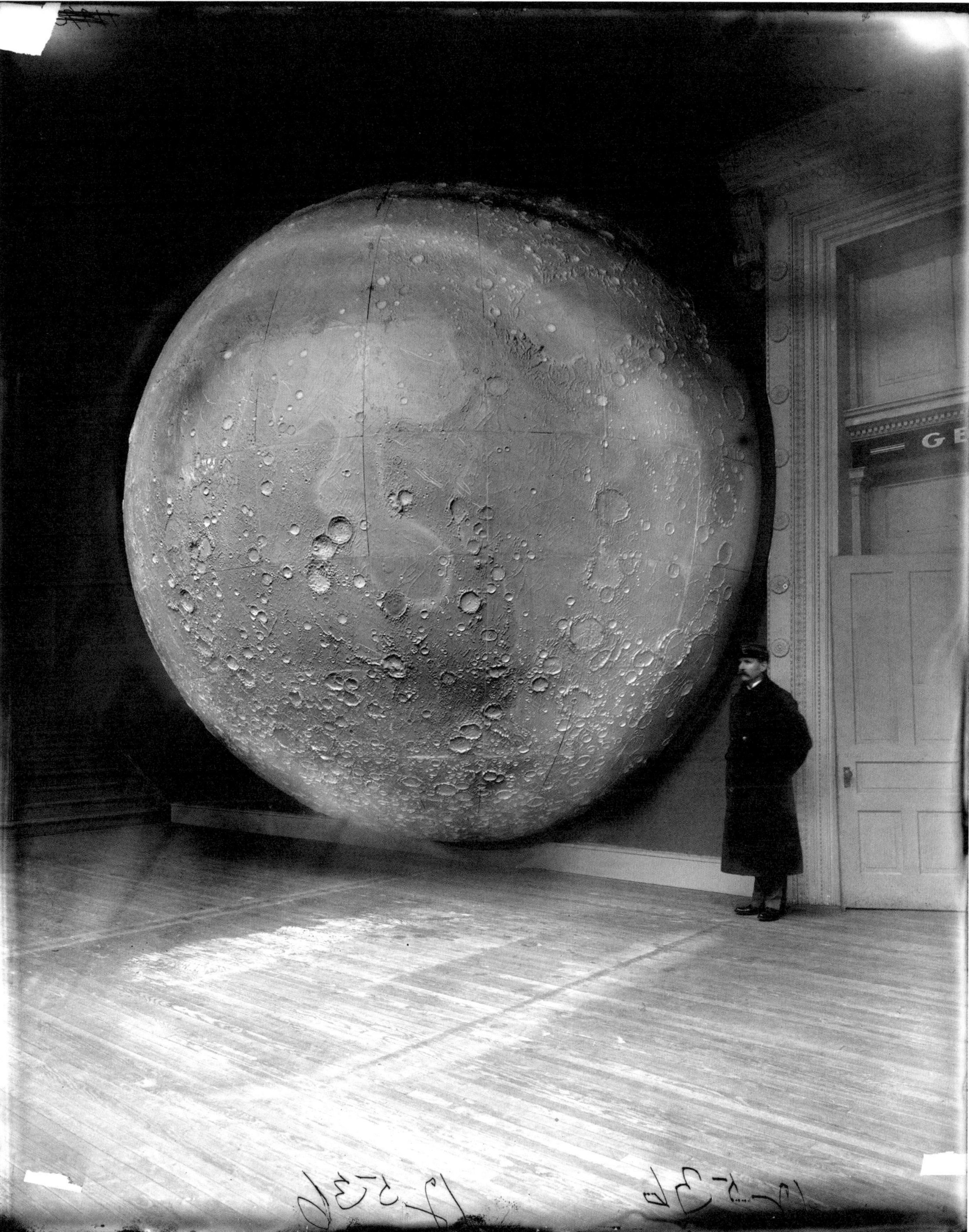

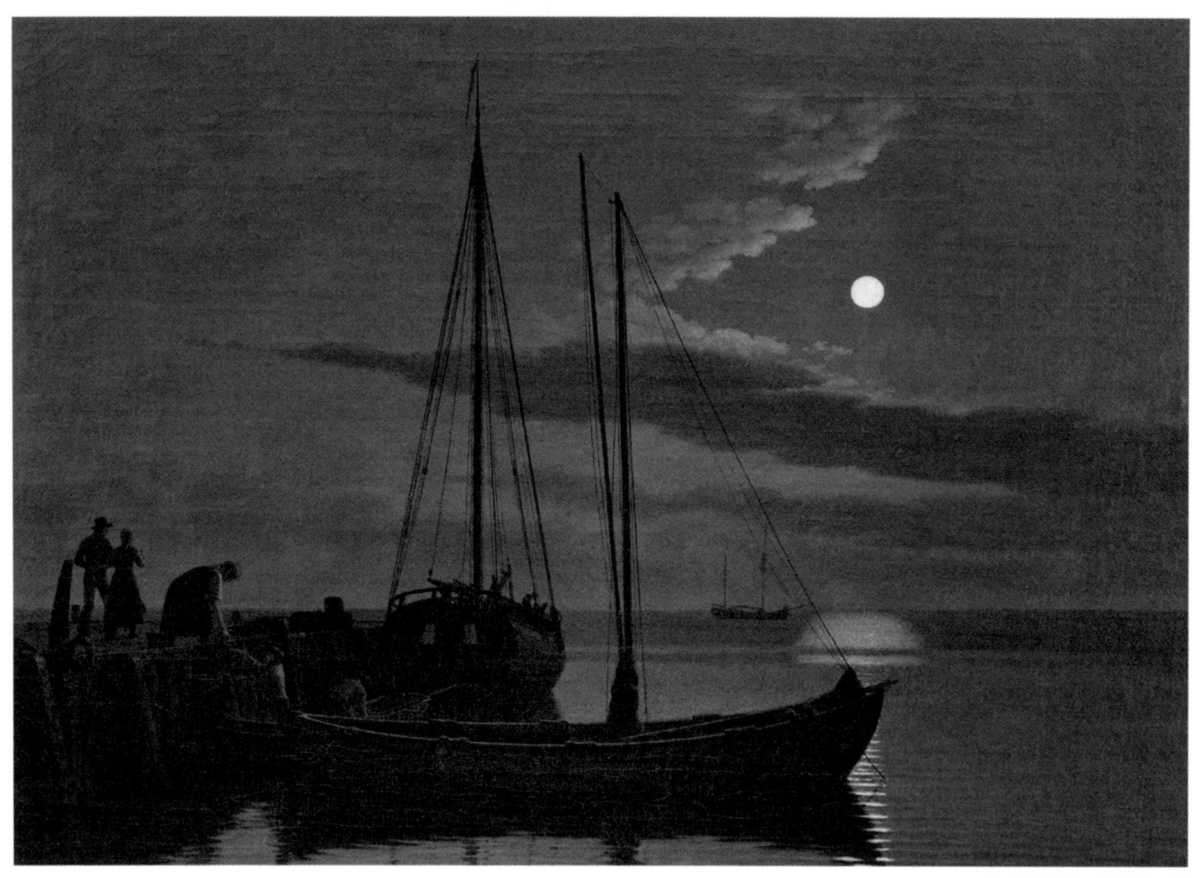

cat. 53
C.W. Eckersberg
Måneskinsbillede, 1821
Moonlight Painting

cat. 108
Katie Paterson
Light Bulb to Simulate Moonlight, 2008

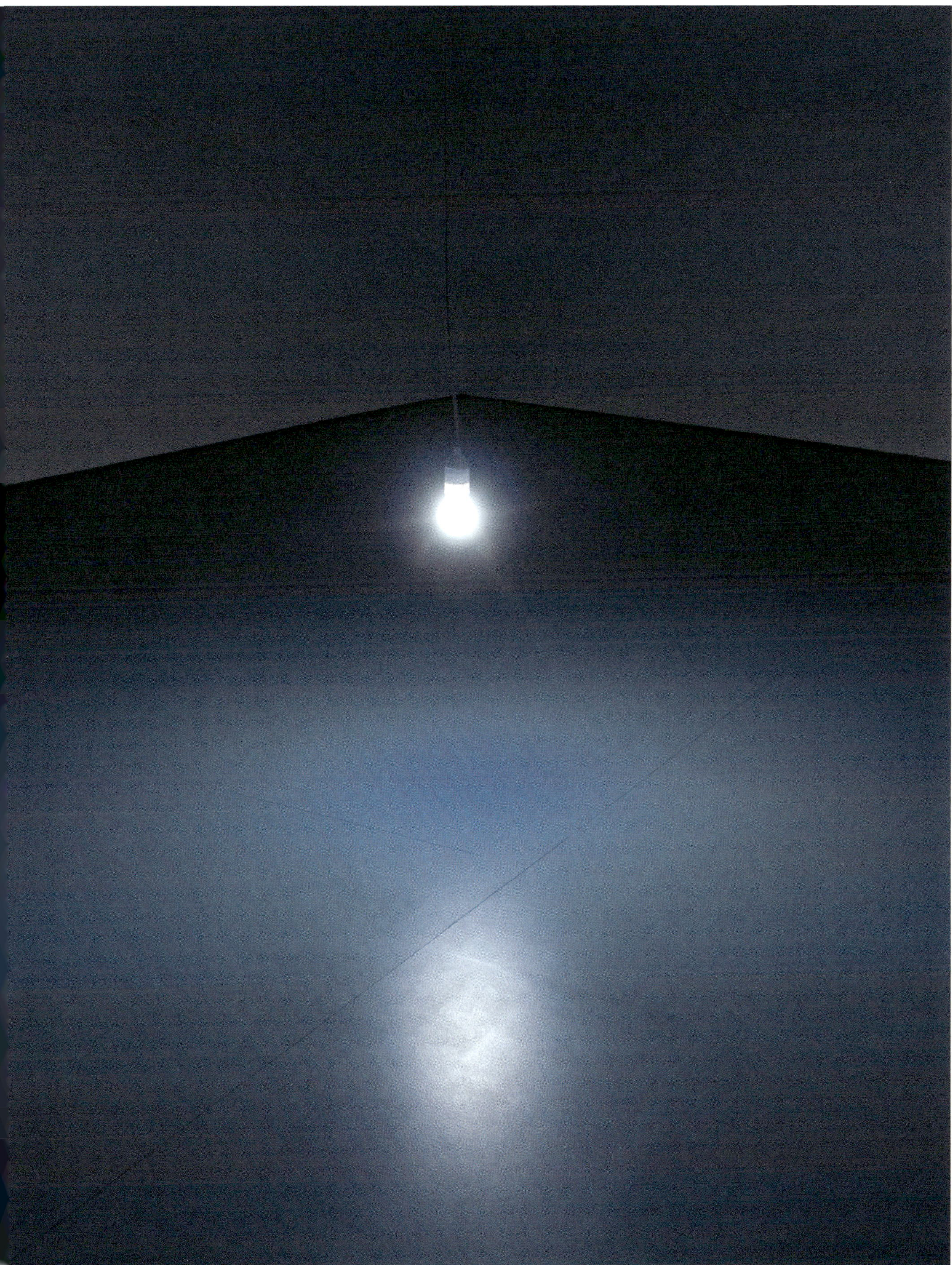

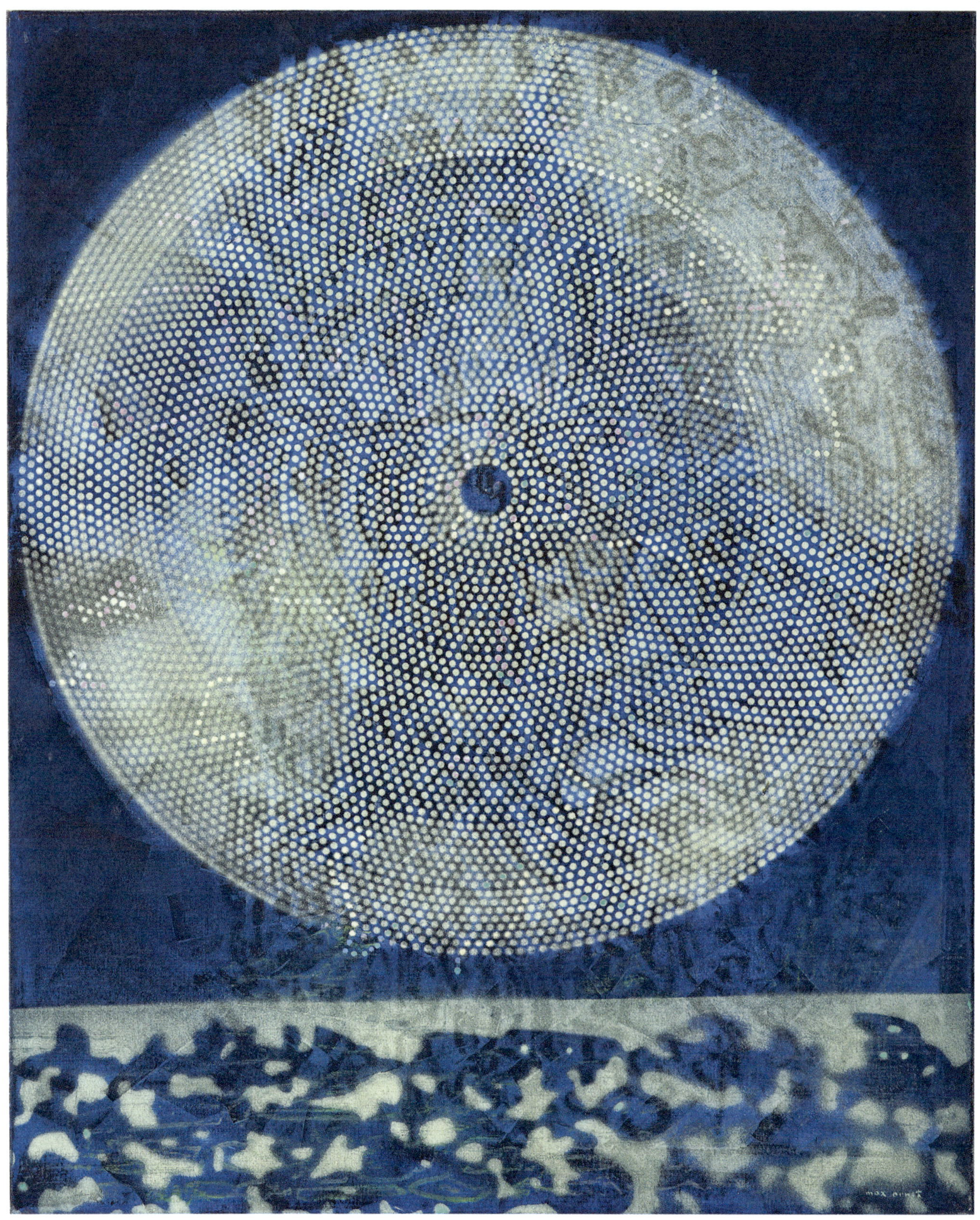

cat. 56
Max Ernst
Naissance d'une galaxie, 1969
The Birth of a Galaxy

cat. 76
Kiki Kogelnik
Fly Me to the Moon, 1963

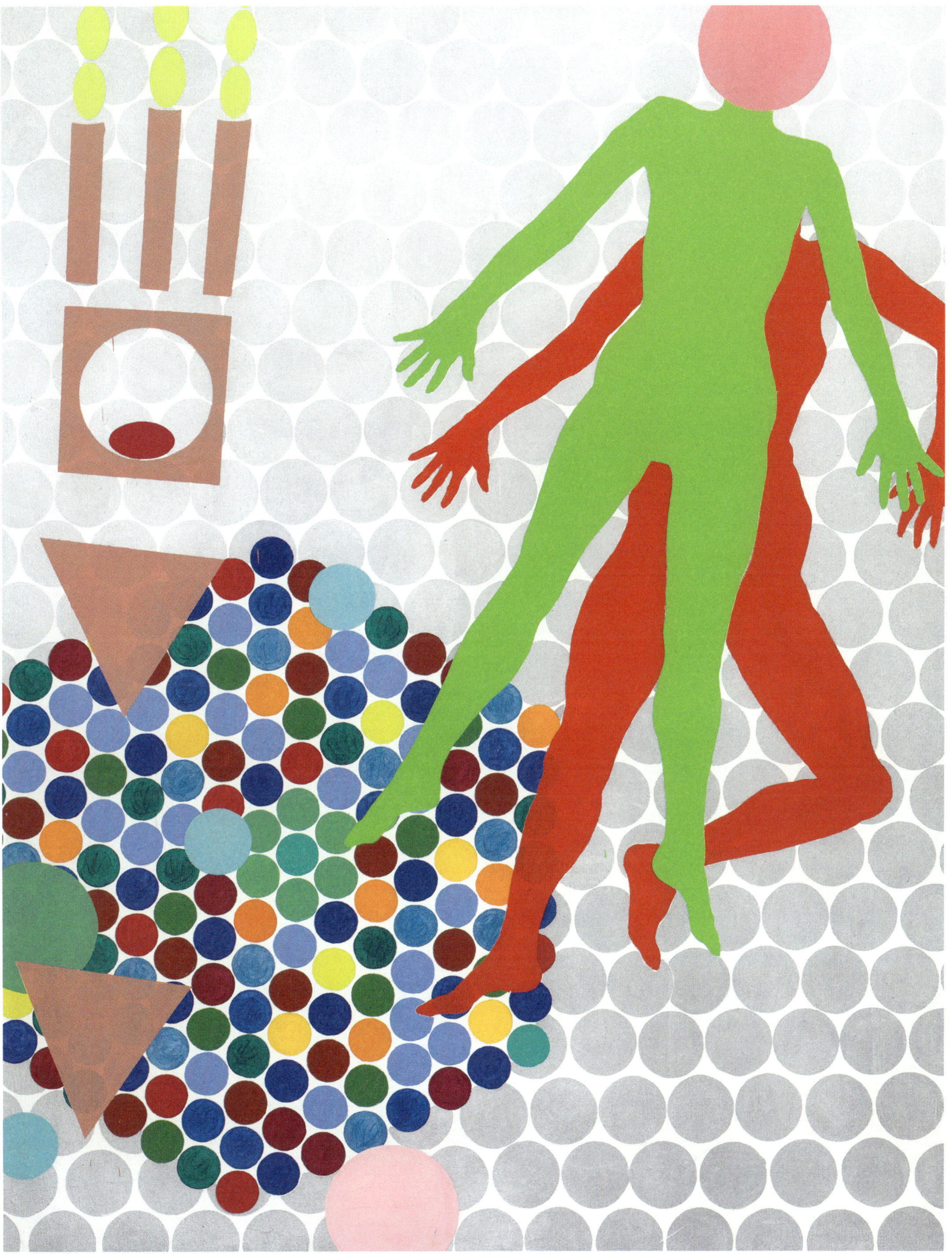

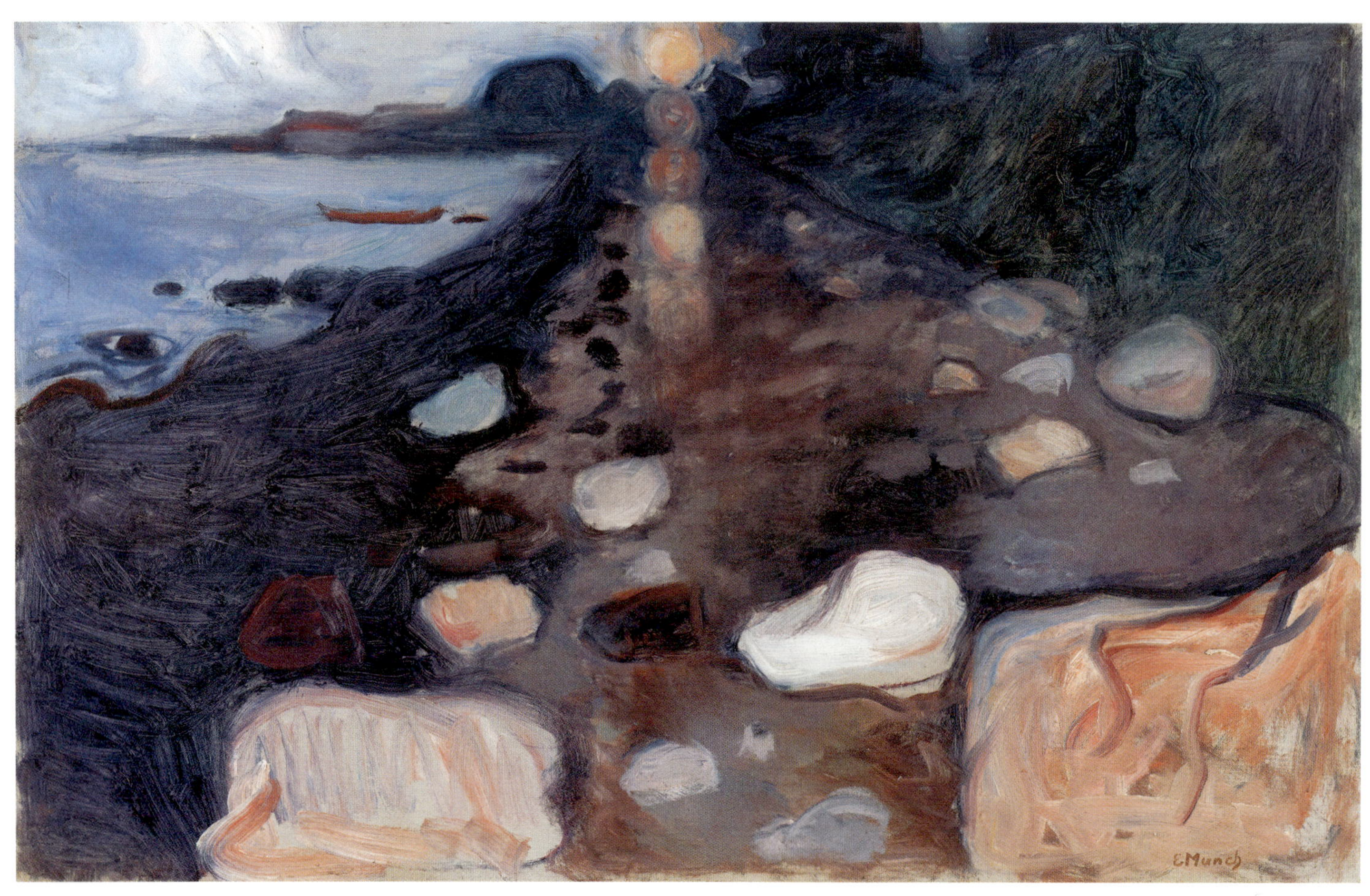

cat. 94
Edvard Munch
Måneskinn ved stranden, 1893
Moonlight at the Beach

cat. 144
Rotraut
Untitled, c. 1972

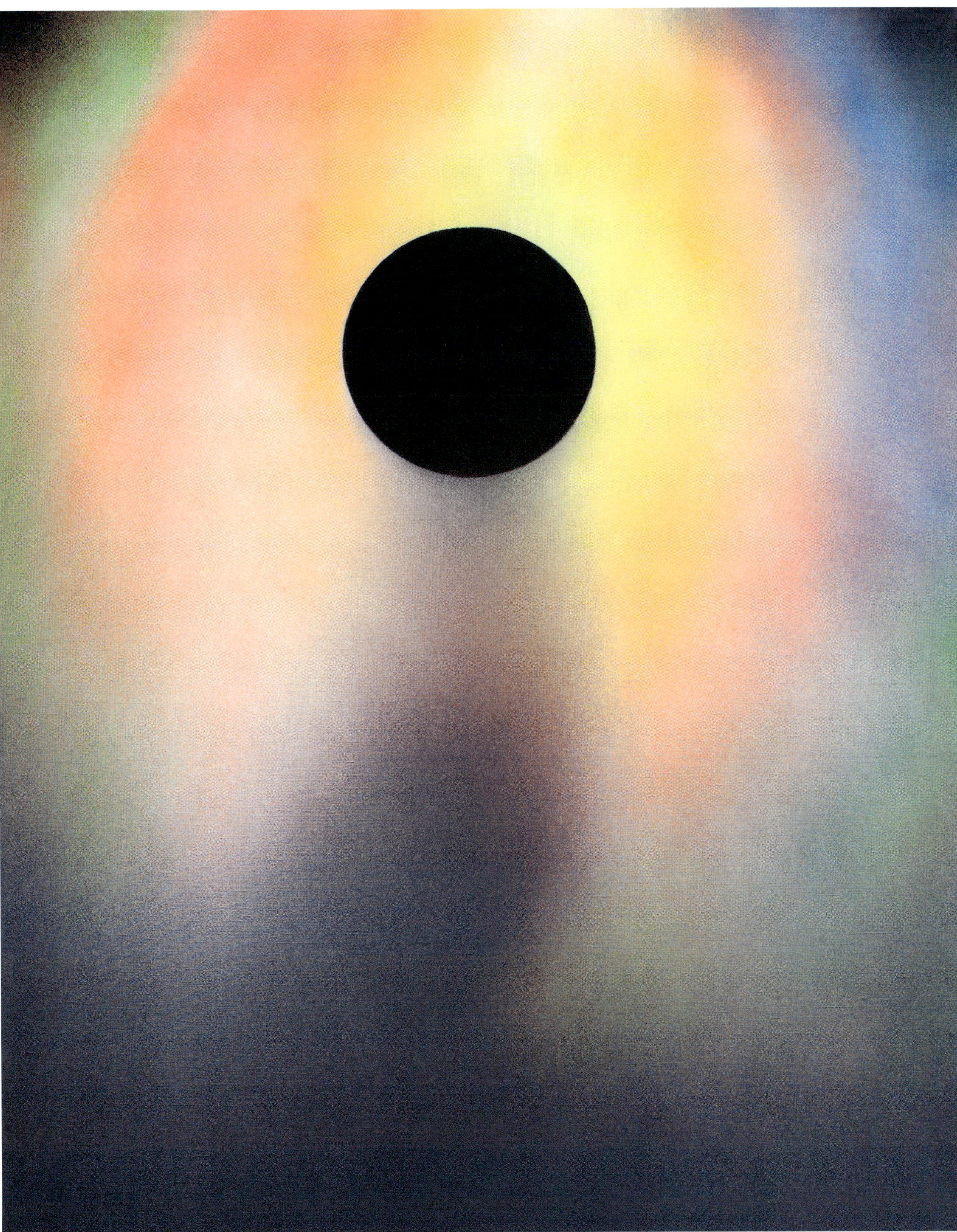

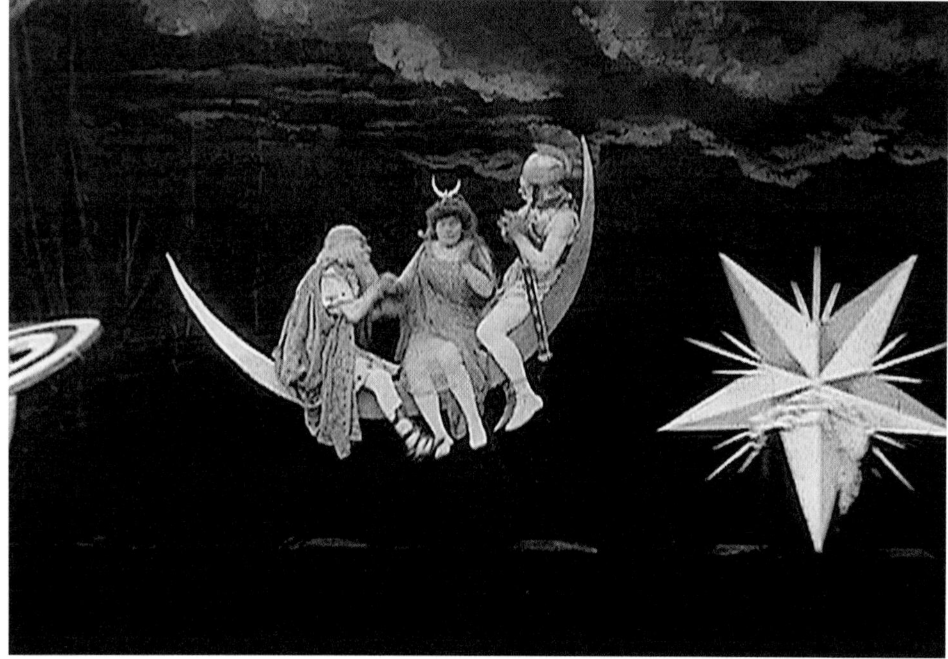

cat. 92
Georges Méliès
Éclipse du Soleil en pleine Lune, 1907
The Eclipse

Acknowledgments

Special thanks to the Aage and Johanne Louis-Hansen Foundation for recognizing the importance of looking at the moon now, in the year leading up to the 50th anniversary of the first manned moon landing, and for generously supporting the exhibition.

Mark Wigley, Dean of Columbia University's Graduate School of Planning
Ferdinand Ahm Kragh
Jacob Lillemose
Stéphane Aquin, Hirshhorn Museum, Washington, D.C.
Jasper Sharp, Kunsthistorisches Museum, Vienna
Anders Toftsgaard, Lene Vinther Andersen and Mette Kia Krabbe Meyer, Royal Library of Denmark
Martin Bizzarro, James Connelly and Christian Mac Ørum, Natural History Museum of Denmark
Sasha Samochina, NASA
Teresa Arcq
Birgitte Agersnap, Thorvaldsen Museum
Martin Appelt, National Museum of Denmark
Tom Watters and Valerie Neal, Smithsonian National Air and Space Museum, Washington, D.C.
Daniel Moquay and Rotraut Klein Moquay
Frances Morris, Director, Tate Modern
Lucy Bamford, Derby Museum

Foreword

"The moon of the nights is not the moon whom the first Adam saw," Jorge Luis Borges writes in a poem. "The long centuries of human vigil have filled her with ancient lament. Look at her. She is your mirror." The moon is an image, the Argentinean poet says. Images, particularly in a museum, are containers of human experience. They are objects for people to look at. When we say that a picture belongs to a certain tradition, we are also thinking about the sum of experiences that people have had with it before us. Pictures are objects of impact history. They meant something to someone before we arrived – we who are here, now. And so, we are not seeing the moon the first Adam saw, even if it is the same one. We see through the filter of history and, perhaps most of all, through the filter of our own experiences. The moon has been a mirror for many things, but the moon is also *my* mirror. And yours. Images are like that. Our gaze occupies them and renews them – on certain conditions, set by big history and by little history, yours and mine.

In art history, the Greek word *topos* is used to describe subjects treated by artists across genres and periods. Many Louisiana visitors will remember the big exhibition we organized a few years ago around another *topos*, the Arctic. The lunar exhibition we are now showing has several features in common with the Arctic one. It likewise paints a big, living, diverse and highly focused portrait of the human fascination with our nearest neighbour in space, the only celestial body that can be seen in full with the naked eye. So far, and yet so tantalizingly near.

There is a wealth of artistic subject matter revolving around the moon. Laid end to end, the subjects tell a story of our mutable relationship to the magical disk of delicate light in the night sky. How our conceptions have changed in modern times can also be read in our images of the moon, which is both a celestial body out there and a place in the inner world of our imagination. It is our mirror, as Borges, who happened to be blind, put it. The moon is a myth machine. There is the Man in the Moon and the moon made of green cheese. The moon plays a part when people get sick or pregnant. A full moon can turn people into werewolves. To name just a few examples. We journeyed to the moon in literature and film, before we actually could. Then, when we were finally able to, we only went a few times. Now, the passion for conquest is heating up again with space colonization and tourism. Yet the moon has not been demythologized. Even in contemporary art, it keeps exerting a powerful pull.

The Louisiana's exhibition, *The Moon: From Inner Worlds to Outer Space*, tells a big story in more than 200 works of art, film, literature, architecture, design and natural and cultural history. We must agree with the poet: the moon is not just the moon the first Adam saw. Unpacking the subject mainly in the time and space of Western culture, the exhibition reveals the moon in its many guises: the mythical and mysterious moon, the poetic moon, the scientific moon, the moon as a destination for science-fiction fantasies and as an altogether more concrete realpolitik-territory for potential possession.

The journey to the moon – from Cape Louisiana – ranges from Galileo's lunar maps to a virtual reality work by the American artist Laurie Anderson, custom-made for the exhibition. Beyond scientific and culture-historical excursions, the trip takes us past Romantic moonlight paintings by Caspar David Friedrich and C.W. Eckersberg, lunar fantasies by Surrealists like Max Ernst and Joseph Cornell and the Space Age aesthetics of modern artists like Robert Rauschenberg and Yves Klein, with a final stop at today's renewed interest in lunar light, as seen in the art of Darren Almond and Katie Paterson. The main weight of the exhibition lies in visual art, but as always when the Louisiana mounts a big production, we make a point of giving the viewer a more inclusive story, beyond the one usually assigned to works of art on museum walls, by mirroring art in cultural history, and vice versa. In many ways, this is an attractive task of today's museums: showing that artworks contain infinitely more than aesthetics, that the artwork is also *your* mirror. Look at it.

Poul Erik Tøjner, director
Louisiana Museum of Modern Art

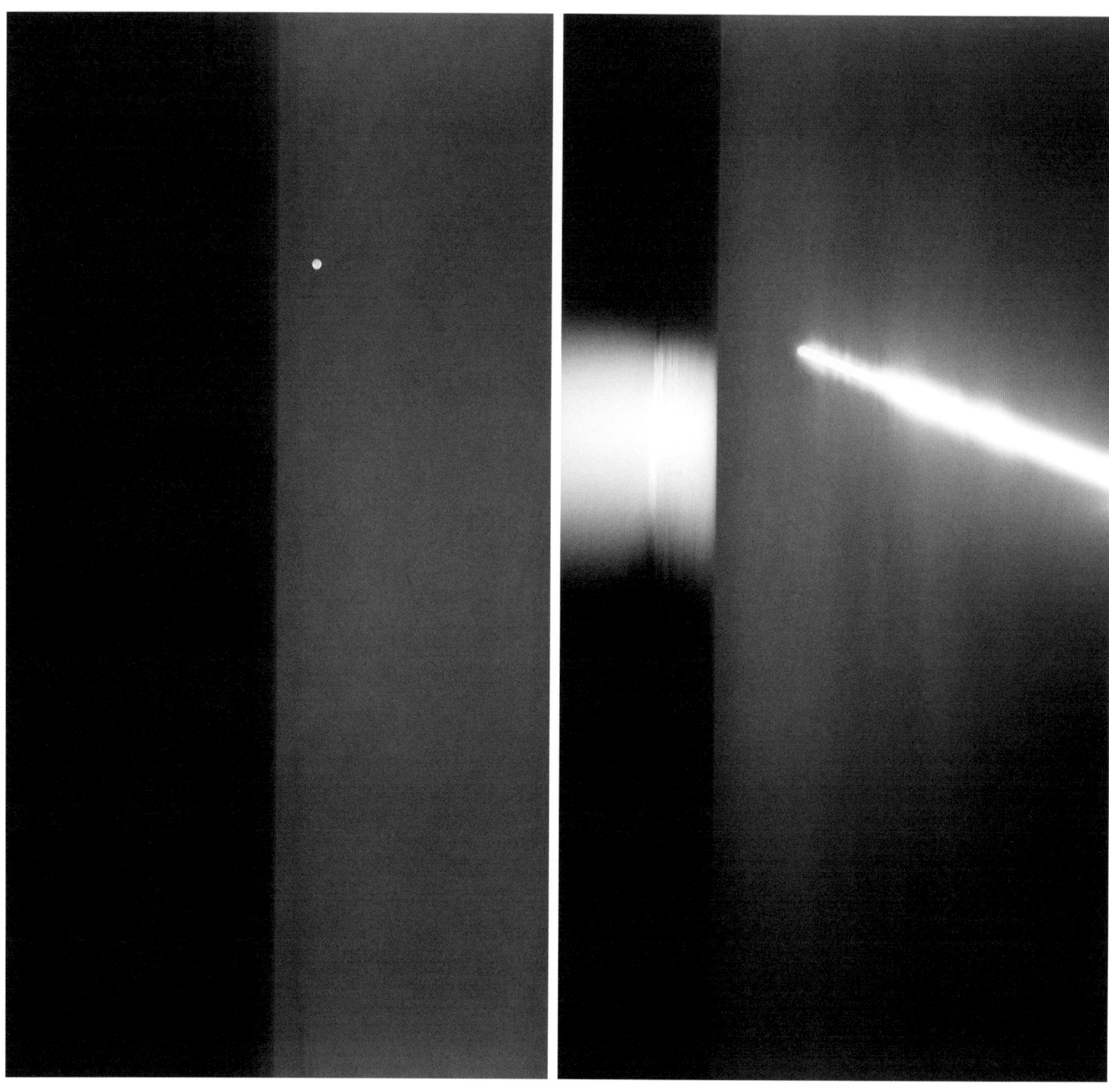

cat. 171
Hiroshi Sugimoto
Caribbean Sea, Yukatan, 1990

cat. 170
Hiroshi Sugimoto
Atlantic Ocean, Newfoundland, 1990

cat. 82
Alicja Kwade
REVOLUTION (Gravitas), 2018

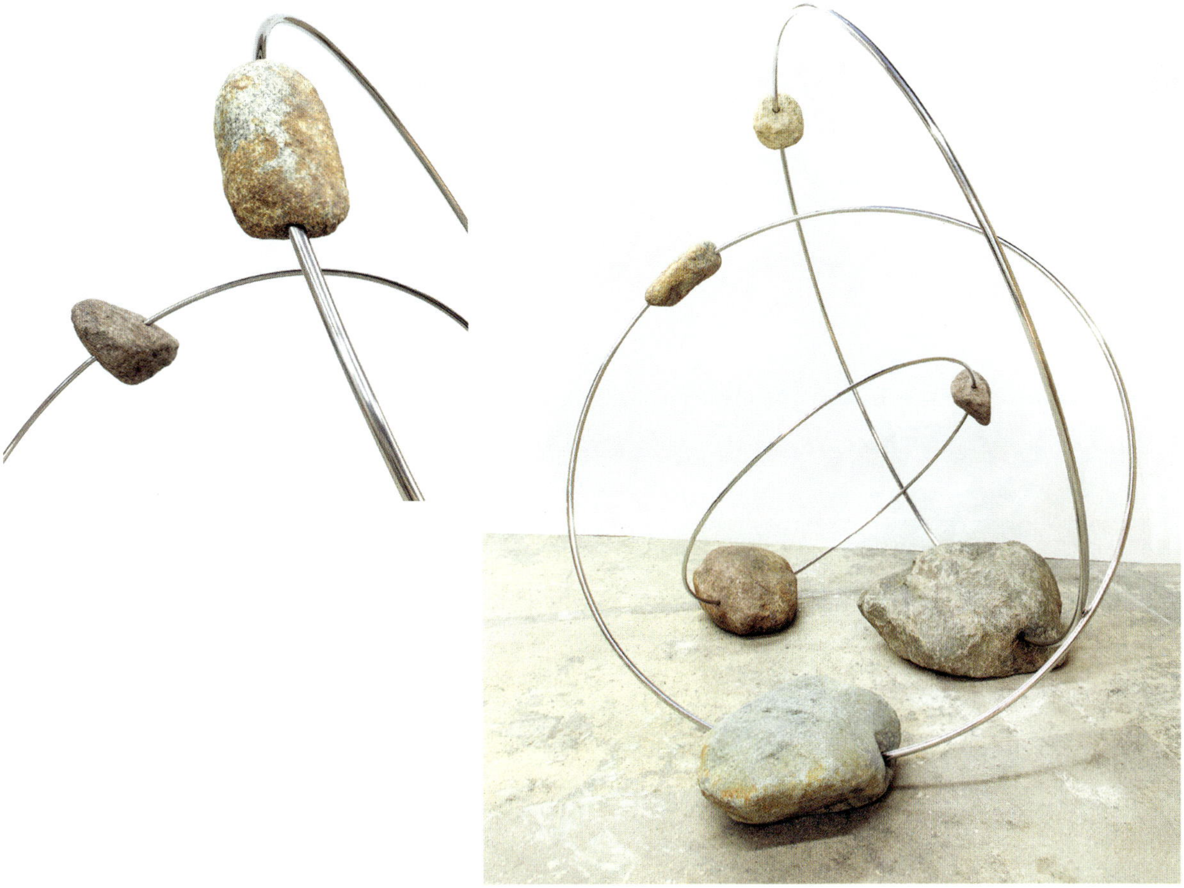

cat. 12
Darren Almond
Fullmoon@Baltic Coastline, 2015

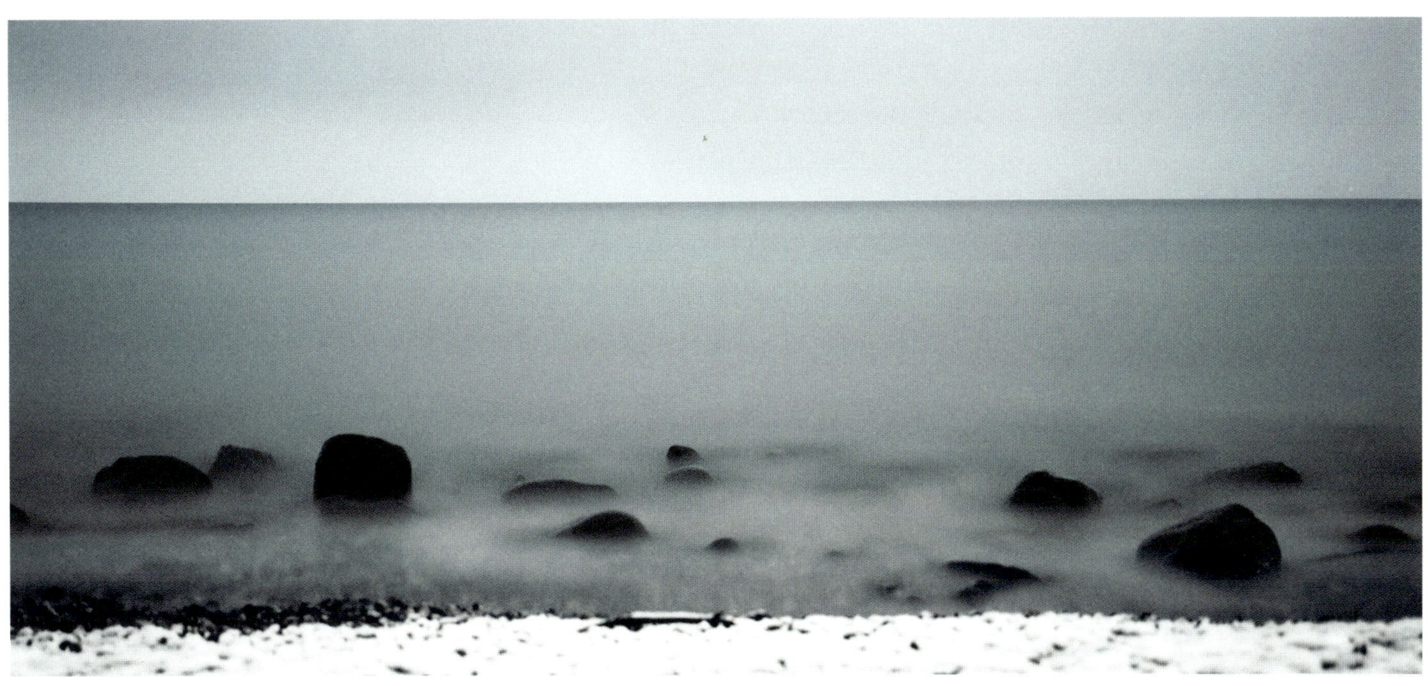

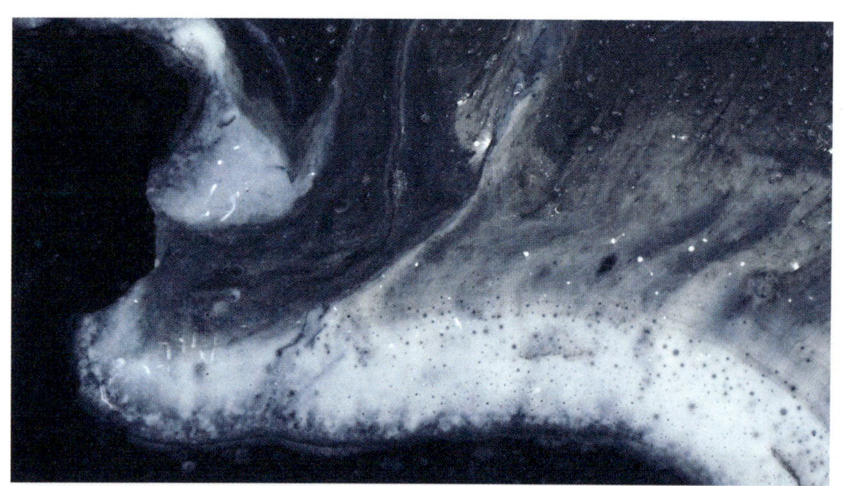

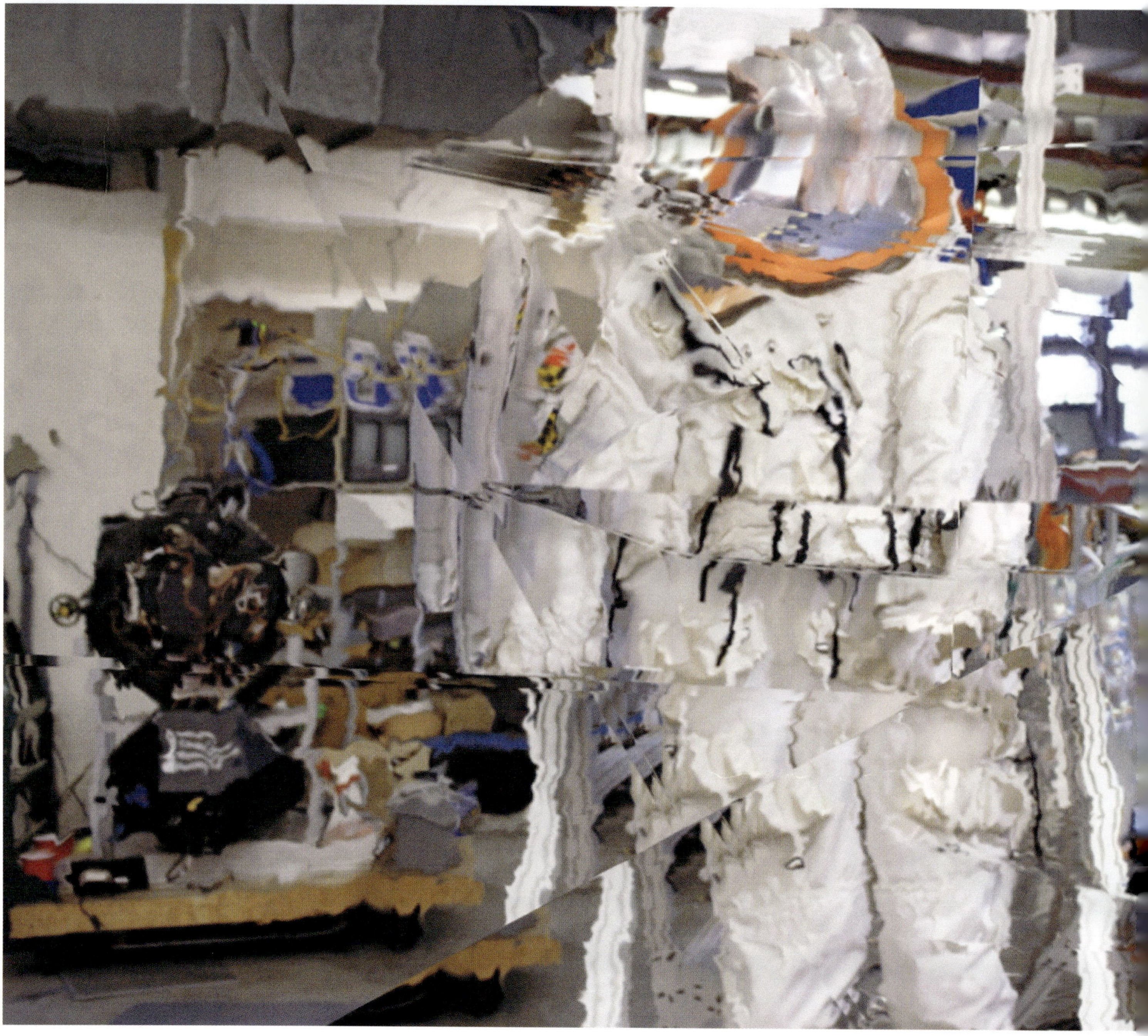

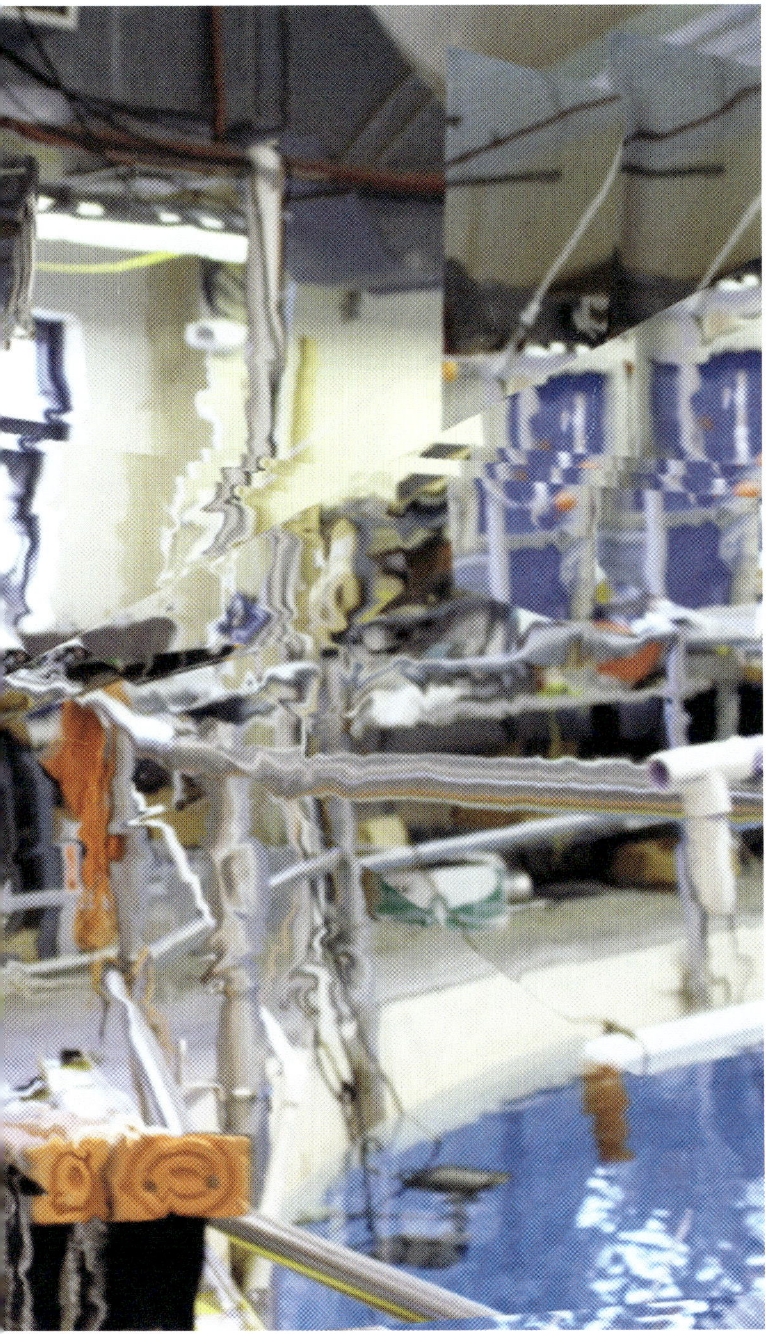

cat. 141
Rachel Rose
Stills from *Everything and More*, 2015

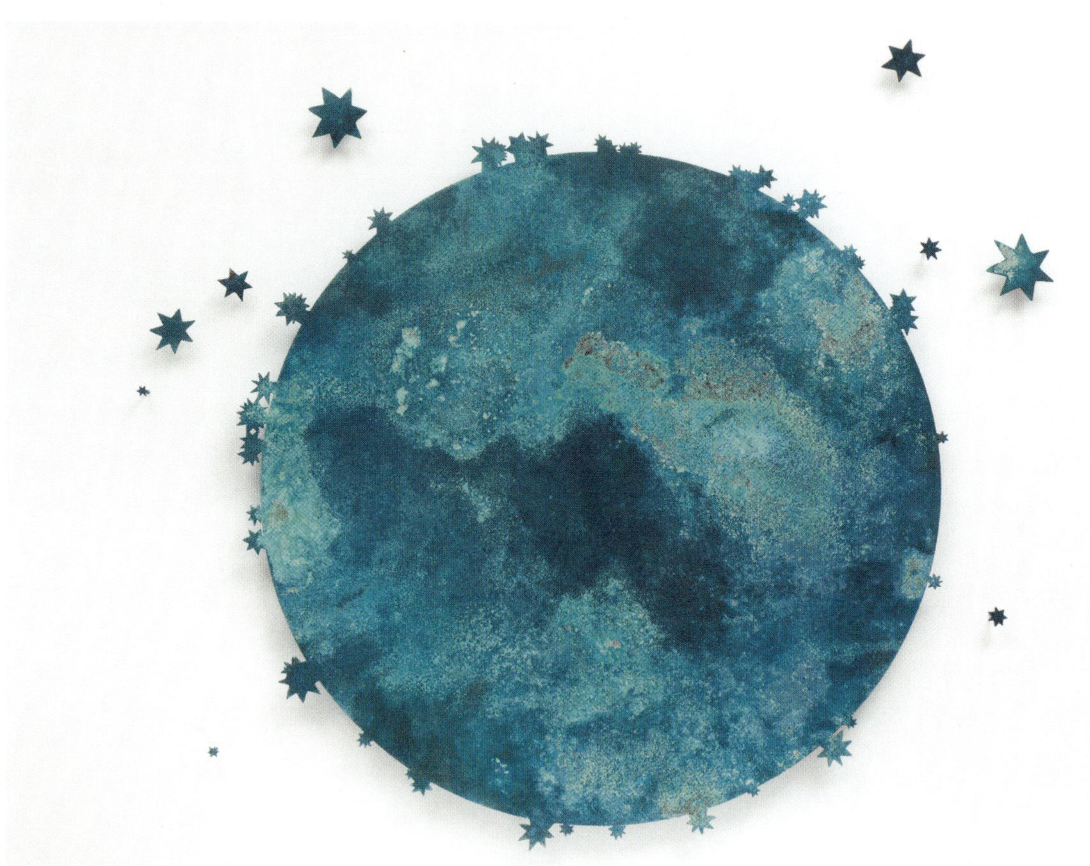

cat. 162
Kiki Smith
Blue Moon III, 2011

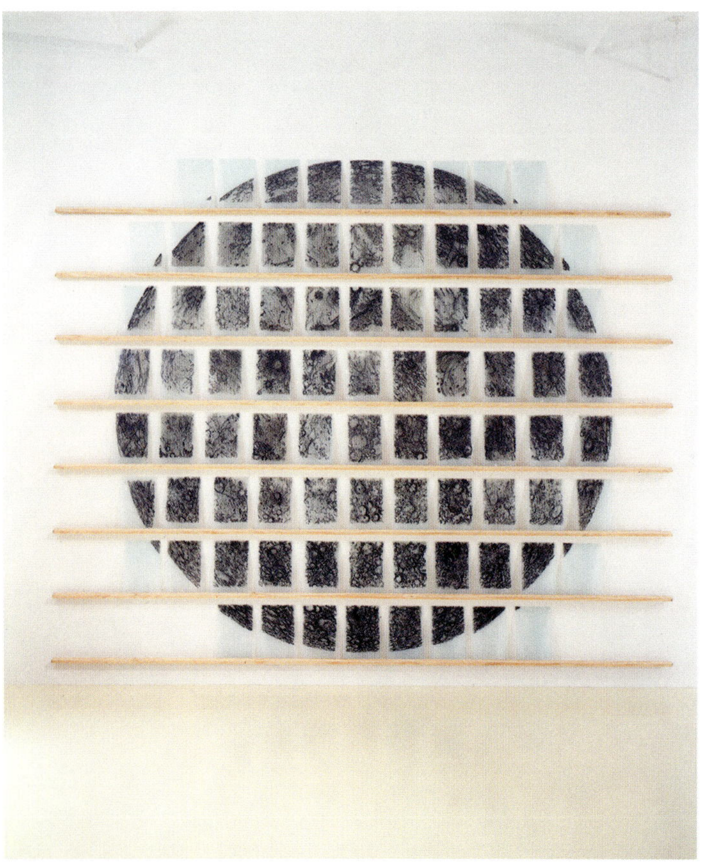

cat. 160
Kiki Smith
Moon, 1997

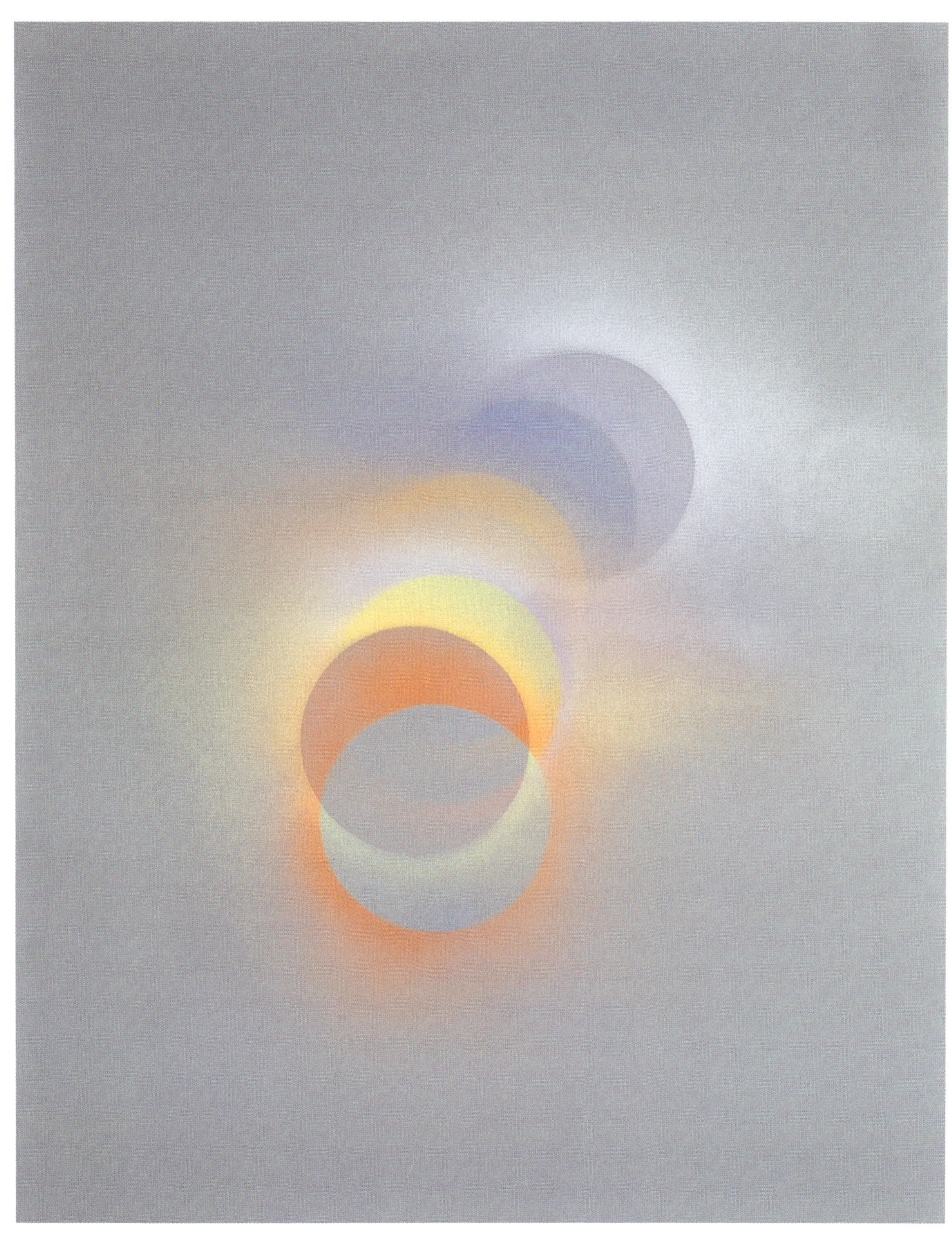

cat. 143
Rotraut
Untitled, c. 1972

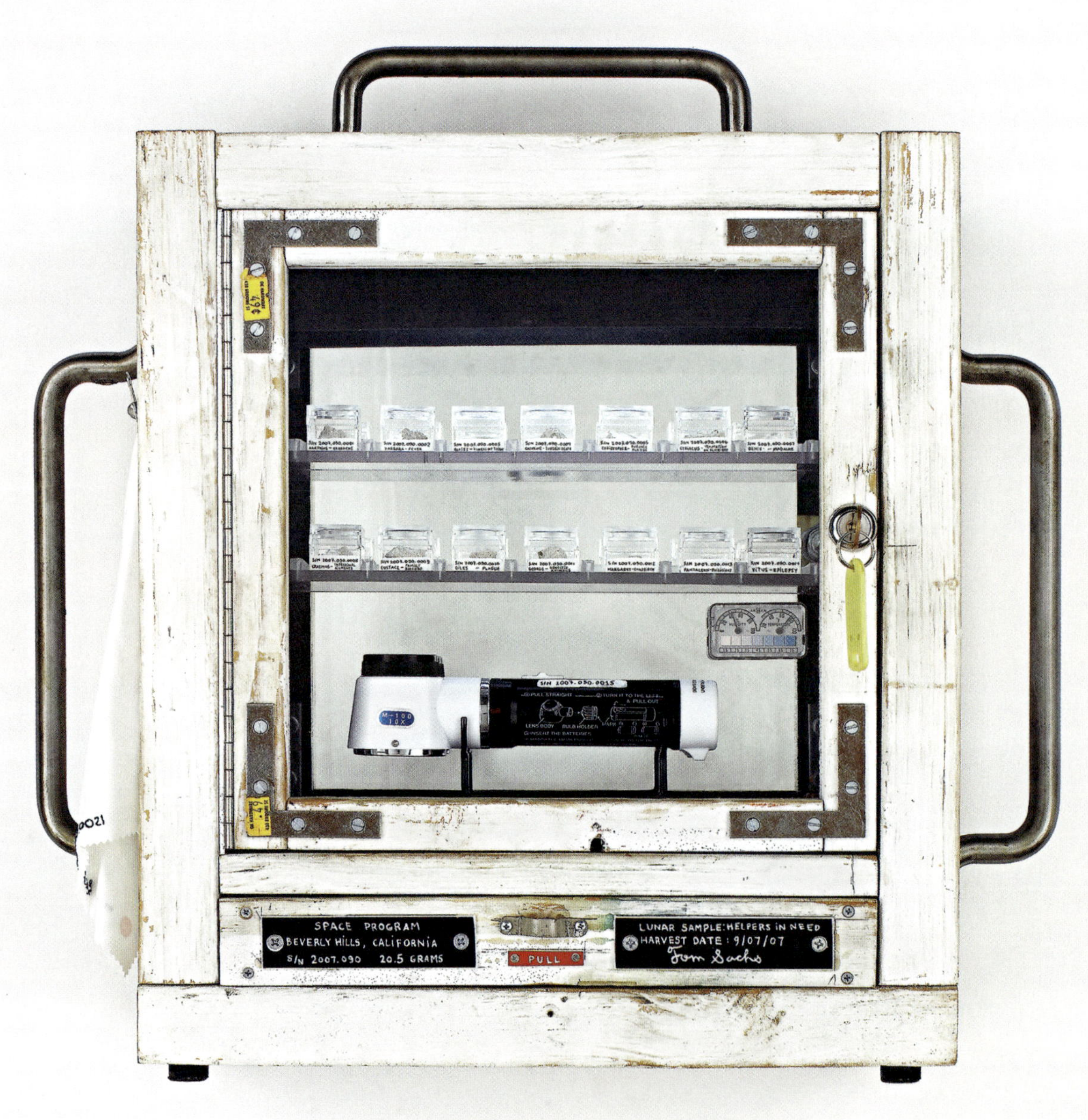

cat. 146
Tom Sachs
Moon Rock Box: Helpers in Need, 2008

cat. 150
Tom Sachs
Europa Rock Cabinet (Sengoku), 2018

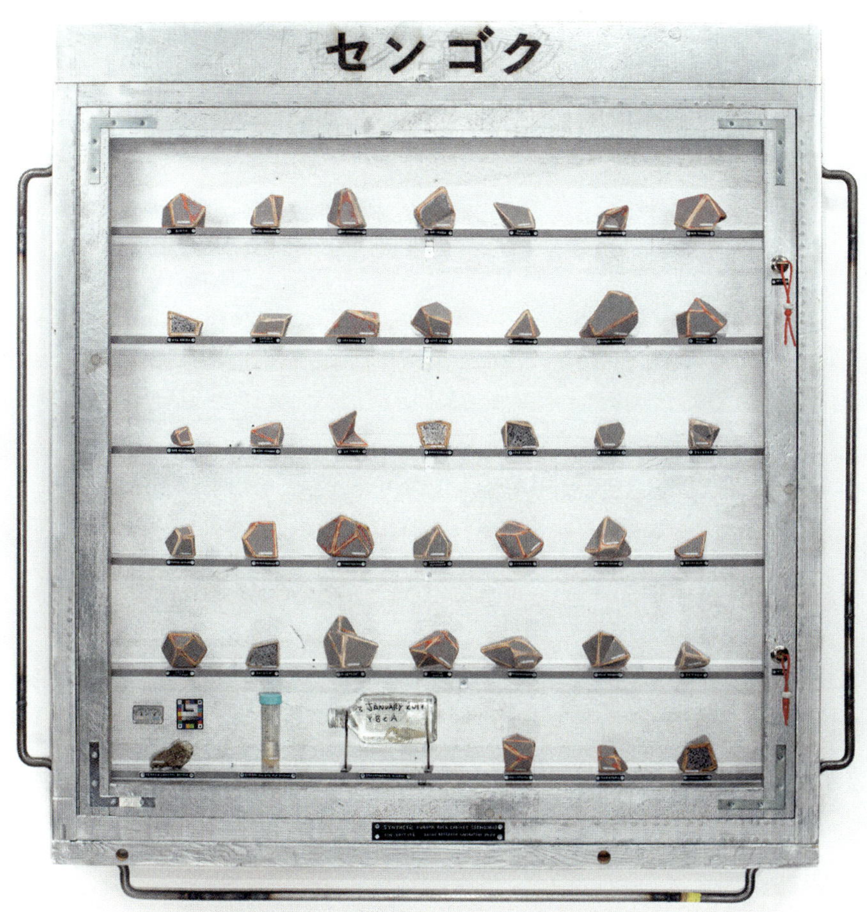

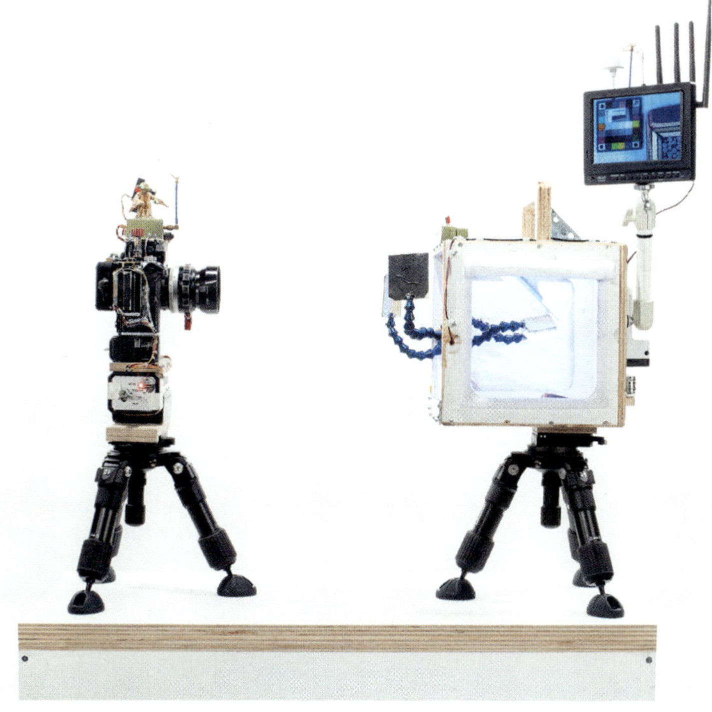

cat. 149
Tom Sachs
Europa Camera System, 2018

The Moon – From Inner Worlds to Outer Space

By Marie Laurberg

Marie Laurberg is a curator at the Louisiana Museum of Modern Art, Denmark, where she has curated exhibitions like *Yayoi Kusama: In Infinity* as well as the present *The Moon – From Inner Worlds to Outer Space.*

What is an exhibition about the moon doing at a modern art museum – here on earth? As the only heavenly body whose surface can be seen with the naked eye from earth, the moon has fascinated artists throughout the ages. Its round white disc has been an open projection screen for myths, imaginings and dreams. The moon is a basic symbol in which the inner world and outer space meet – science and folklore, fiction and technology, existential searching and the urge towards economic expansion. And in visual art these ideas are given condensed expression, where the moon is a mirror for mankind's thoughts about existence and our place in the world. What is important to us, to what do we ascribe value, how do we create meaning in the world? Ask the moon. It hovers in the sky like a blank white screen and shows us the stories we tell about it.

It is the moon as a mirror of culture that we investigate in the Louisiana's exhibition *The Moon – From Inner Worlds to Outer Space.* In the cultural-history perspective of the exhibition the moon raises questions about the meanings that can be ascribed to our floating around in an infinite universe. The moon offers us a collective moment of introspection and reflection over our fragile existence and the place of mankind in the universe. Generation after generation, rotation after rotation.

The exhibition paints a multi-faceted portrait of the role of the moon in the Western imagination with more than 150 works and objects – art,

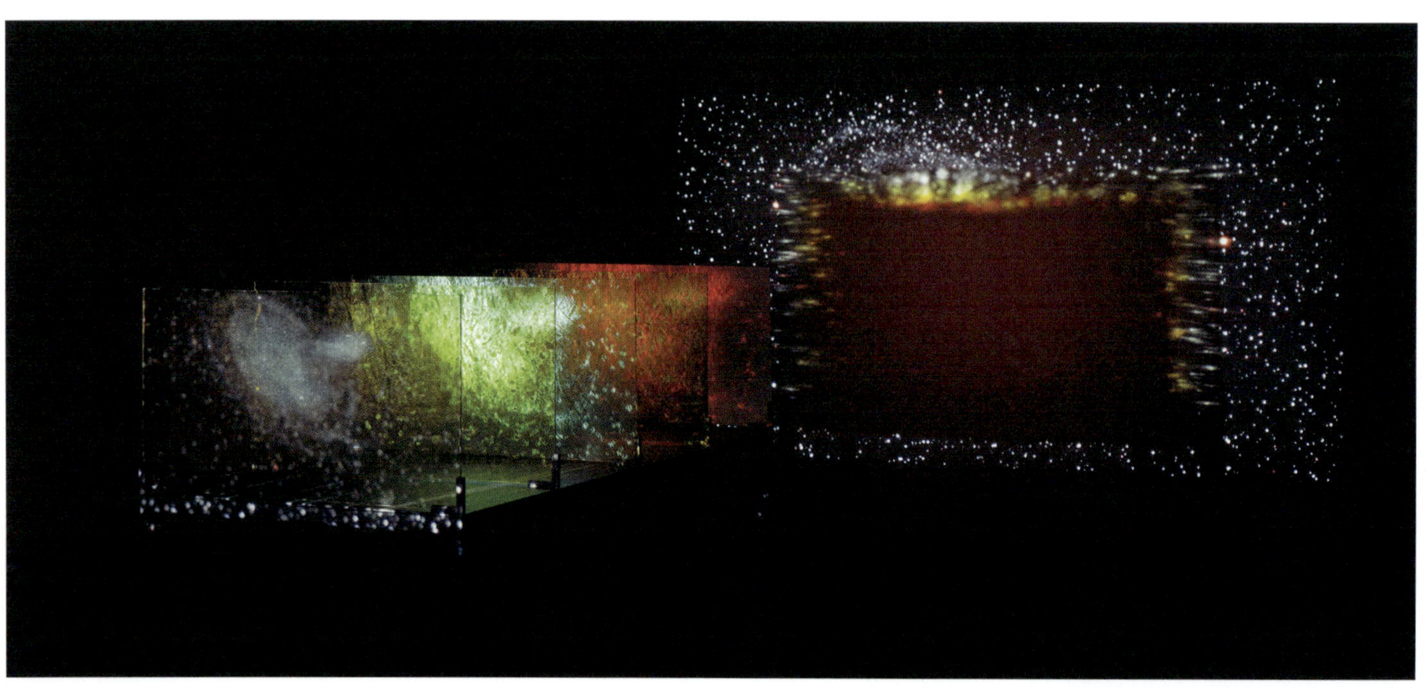

cat. 24
Rosa Barba
The Color Out of Space, 2015

literature, music, film, design, cultural history, architecture and science – collected over several years of intensive research. The aim of the exhibition's interdisciplinary perspective is to engage objects from different fields of knowledge in direct dialogue and thus underscore the role of art as a special kind of knowledge – a means of exploring the world. Art does not develop in a vacuum. It develops in dialogue with the knowledge that arises within other areas, of which the exhibition has many examples: American Robert Rauschenberg was invited by NASA to witness the Apollo 11 launch. Galileo's pioneering map of the moon was created with one foot in the world of science and the other in that of art. One of the nineteenth century's principal astronomical works on the moon was made by a researcher, James Nasmyth, who was also an artist. And the Surrealists incorporated both astronomical science and the mythological and religious objects of other cultures in their images of the moon to make contact with the deepest spiritual layers of humanity.

With dialogues across genres the exhibition is organized as a number of thematic nodes where the moon has a particular, defined place in the world of cultural ideas. In some periods the moon has shone more clearly through the history of art than others: the pictures of Romanticism are often bathed in atmospheric moonlight and speak of the smallness of mankind in the face of the greatness of nature. The pictures of Surrealism are concerned with the moon as a power that exerts a pull on mankind. Hardly surprisingly, the moon again became an important motif in art during the space race of the Cold War – as a monument to mankind's mastery of the cosmos. And in contemporary art we meet a critical gaze at the effects derived from the achievements of modernity: an interest in the colonization of space, which goes hand in hand with ecological crises and an eye for the deep cosmic layers of time that transcend the human.

The Planetary Perspective

Contemporary art's representations of the moon are tied to a broader interest in outer space and the properties of the earth as a planet. I will call this tendency a *planetary awareness* in contemporary art. Themes like geological conditions, fundamental laws of nature, microorganisms, fossils, the atmosphere, plate tectonics, the solar system and its celestial bodies, and the mysterious dark matter of the universe, appear in artworks which approach nature in a new way: not as a landscape, a political territory or a post-modern cliché, but as a cosmic phenomenon with more or less clear references to the current ecological crisis, directing our attention to our planet and its destiny viewed in a long-term perspective. The works take their starting point in the fact that the fate of mankind is inextricably bound up with the planet earth. Our civilizational activities are played out against the background of a far greater geological time-frame, the 'deep time' that is not measured in hours, weeks and years, but in the lives of planets, stars and galaxies. For in the deep time of the universe, humanity is nothing but a blip.

As the astrophysicist Anja C. Andersen says elsewhere in this catalogue (p. 108), the genesis of the earth and the moon are one. They came into being at the same time as a result of a huge collision in the early solar system around 4.5 billion years ago – a dizzying timespan. The philosophical concept 'deep time' is applied to this timescale, and was introduced in the eighteenth century by the Scottish geologist James Hutton. If we zoom out to this deep timescale, mankind almost becomes invisible. One of the contemporary artworks that deal with deep time is the film installation *The Colour Out of Space* (2015) by the artist Rosa Barba. The work projects images of the moon, stars and planets photographed from the Hirsch Observatory through discs of coloured glass, so they are refracted and reflected. On the soundtrack of the work, researchers, artists and writers reflect over the endlessness of the universe – aspects such as the strange displacement of perceived time by the speed of light, which means that what we see when we look deep into the universe is images of stars and planets as they looked millions of years ago; or the fact that modern image technologies are the precondition for being able at all to transform what is 'out there' into visual forms that the human eye can capture. There are many filters and layers of images which, like the glass discs in Barba's work, interpose themselves between us and the vast cosmos out there. Even the latest images captured by the Hubble telescope are based on interpretation and fictionalization – they are shaped, coloured and named: the Eagle Nebula, the Sombrero Galaxy, the Cat's Eye Nebula. This shows how image formation is a fundamental tool in mankind's attempt to bring the cosmos within reach – a condition that applies to science as well as art. In this way the two forms of knowledge, the scientific and the artistic, are more closely connected than one might think. They both makes use of images and image formation as a way of understanding the world.

cat. 172
Hiroshi Sugimoto
Red Sea, Safaga, 1992

The image is precisely that – a tool of understanding – for the artist Hiroshi Sugimoto, whose black-and-white photography deflects the medium away from its traditional connection with documentarism and instead visualizes fundamental principles such as time and space, past and present. The moon over the sea is the basic motif of the photographic series *Revolution* (1986-1997), where the horizon line is tilted vertically and divides the image sharply into sky and sea. In some works the moon draws a thick strip of light across the black sky, so we sense its orbit around the earth in an uninterrupted motion. The light of the moon is reflected in the metallic surface of the sea. Sugimoto has stated that the point of departure for the *Revolution* series was an experience of nature that made him aware of mankind's place in a larger universe. Sugimoto refers to astronomers like Copernicus and Galileo, whose observations of the planets in the firmament instituted a gaze at the world detached from the earth. The photographs have a strange, supraterrestrial effect that gives the illusion of a "view from nowhere." Their vertical horizon gives no indication of where we stand, there is no firm ground beneath our feet. In front of the pictures we seem to float before an enormous expanse of space with no sign of life. The world seems huge, undisturbed and inscrutable – as if we are looking into a world before the advent of human beings.[1]

The scope of the exhibition has also been broadened to include works which deal in a more abstract way deal with the cosmos and man's place in the universe. Alicja Kwade's sculptures work with basic cosmic forms in a more direct, bodily way: granite rock is apparently thrown around among steel rings in dynamic planetary orbits (p. 13), and large globes of coloured marble lie as heavy, physical images of celestial bodies. The artist herself has compared the work to a goddess throwing the planets of the solar system as in a game of marbles. Yayoi Kusama's large installation *Gleaming Lights of the Souls* (2008) from Louisiana's own collection lets us travel out into the universe for a while: here we step into a small, mirror-like space where coloured lamps like planets or stars flash and reflect endlessly. Perdition and self-annihilation, abandonment to a cosmos orchestrated by small electric lamps.

Moonlight

When did you last experience moonlight – undisturbed by other sources of light? Moonlight in its pure form is impossible to find in a city, and the increasing light contamination over large parts of the planet means that nature conservation groups have in recent years been calling the night sky a phenomenon worth preserving. Katie Paterson's work *Light Bulb to Simulate Moonlight* (2008, p. 5) offers its own solution to this issue – urban man's paradoxical attempt to maintain a particular experience of nature. Paterson has collaborated with the light-bulb firm Osram and a group of engineers in making light measurements, analysing wavelengths and finally identifying a suitable surface treatment so that the light quality of the specially manufactured light bulbs simulates

cat. 85
Carl Julius Leypold
Kirchhofseingang, 1832
Entrance to the Cemetery

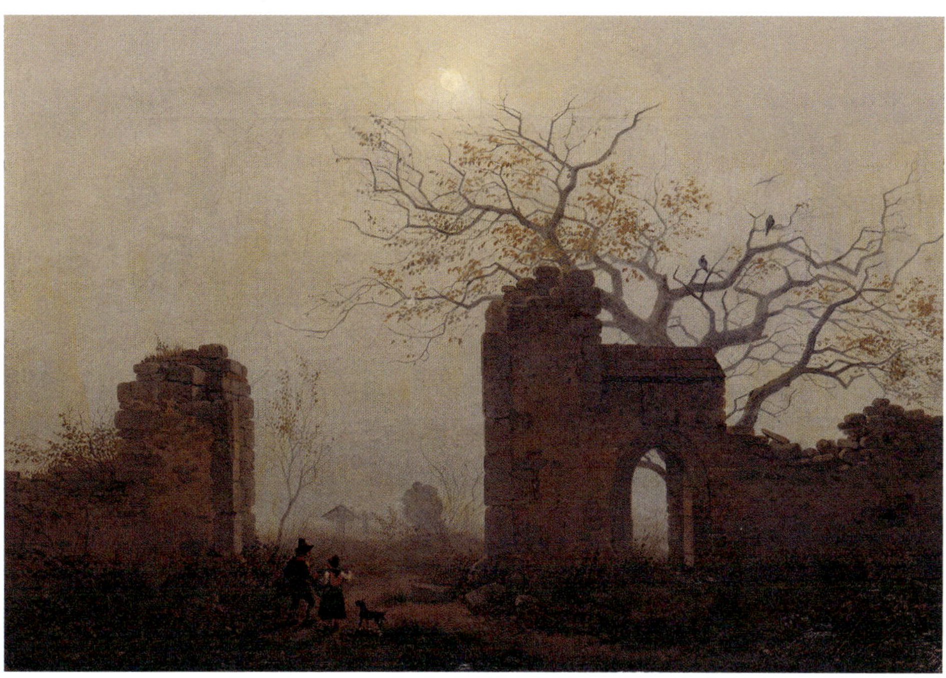

moonlight as accurately as possible. Each of the 289 light bulbs in the work has a lifetime of 2,000 hours, so with a total duration of 66 years this corresponds to an average human lifespan in 2008. A life-long consumption of light. Every time the work is exhibited, the whole collection of bulbs is shown, but only a single one is in use, and bathes the room in its characteristic, cold light. When one bulb goes out, a new one is switched on. The light gradually runs out and a human life is measured out. The work is characteristic of Paterson's use of everyday objects to make a transient phenomenon like human mortality tangible: the duration of life measured out in light bulbs that burn out.[2]

Paterson's work is a present-day updating of one of the fundamental motifs of Romantic art: moonlight as an occasion for contemplation. For centuries the special light of the moon has been a source of fascination for artists and writers. The distinctive quality of its light – which is in reality a reflection of the light of the sun – is perceived as bluish by the human eye. By the sparing light of the moon the sensitivity of the eye to details and colours is reduced, and a greyish, dim world emerges from the darkness of the surroundings. Looking through the European landscape painting of the 1800s we find so many moonlight studies that we can speak of a regular genre. Although the moon often appears as a motif in the pictures, it is not in focus as an object of study – it is its cool, contour-erasing light that attunes the mind to a particular mood. The moon offers an opportunity to contemplate sublime nature, which with its mighty, eternal power stands in contrast to the ephemerality of human life. To find the foundation of contemporary art's planetary awareness we must thus go back to Romantic art, where the moon plays a major role.

The painter Caspar David Friedrich is a crucial figure in German Romanticism. For Friedrich the night sky with its constantly shifting mysterious light was a divine phenomenon.[3] He was particularly interested in dawn and twilight, in sunrise and sunset colours, in fog – and moonlight.
In the painting *Seashore by Moonlight* (1818, p. 35) Friedrich shows the horizon line by the sea bathed in moonlight. Fishing nets draped on thin stakes stuck into the sea bed are hanging in the foreground, their branches reaching up to the sky like stiffened fingers. A frail construct among the large, heavy stones of the coastline. The diffuse light of the moon bathes sky and sea. Eternity and mortality meet here at the fishing hamlet.
And it is this very role that moonlight plays in the painting of the period – as a mediator that puts small mankind in contact with Great Nature. In the work of Friedrich's contemporary, Carl Julius von Leypold, human mortality and the eternal light of the moon meet at the entrance to a churchyard in the painting *Entrance to the Cemetery* (1832). The human figures in the painting, a woman and a man accompanied by a small dog, move among a whole array of mortality motifs: ruins, graves and half-dead oak trees form the backdrop for a walk which in the scenic space of the picture becomes the

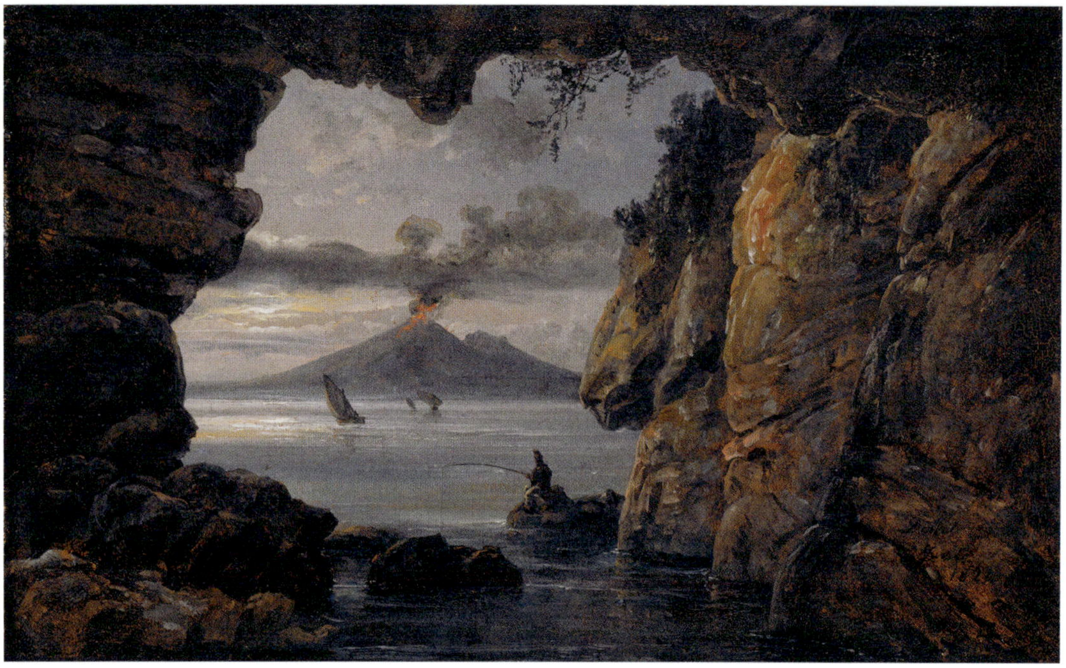

cat. 42
J.C. Dahl
The Bay of Naples seen from a Grotto, 1821

journey of life. With its cycle of phases – from new moon to full moon – the moon is in itself a symbol associated with both new life and death – and in Romantic art this life cycle is measured against sublime, almighty nature.

The moon played a crucial role in the culture of the early nineteenth century, and the art historian Sabine Rewald speaks of a decided "lunar period" in the German culture of the time.[4] The phenomenon is not limited, however, to Germany; it can be observed in art over large parts of Europe, including France, Scandinavia and Britain. The German philosopher Arthur Schopenhauer offers his own explanation of the fascinating power of the moon: "it is *sublime*, i.e., sets us into a sublime mood, because, without any reference to us, eternally foreign to earthly doings, it moves on and sees all but is involved in nothing... In consequence of this whole beneficial impression on our disposition, the moon gradually becomes our bosom friend, which to the contrary the sun never does; like an exuberant benefactor whom we are completely incapable of looking in the face."[5] The sublime power of the moon lies in its distance from us. It is at once accessible and visible, but still always at a distance, unattainable. Several cultural historians have pointed out that Romanticism's view of nature was a reaction to the period's scientific discoveries and technological advances – or what the sociologist Max Weber has described as modern, secular society's "disenchantment" of the world;[6] a process of rationalization where the mysticism and spirituality of traditional society has been phased out in favour of a scientific and rationalized worldview.

The light of the moon was also a pivotal point in a succession of compositions from 1800s. Ludwig van Beethoven's *Moonlight Sonata* (1801) already gained great popularity in its own time and was given its name five years after Beethoven's death, when the poet and music critic Ludwig Rellstab compared the introductory sequence to the effect of moonlight on Lake Lucerne.[7] In the exhibition this piece is a focal point for a contemporary artwork by the above-mentioned Katie Paterson, in which the score of the *Moonlight Sonata* is transmitted in Morse code to the moon, after which its is reflected back to the earth. Certain parts disappear into the moon's craters, and the returned piece of music with all its gaps is played on a grand piano.

The Romantic Moon

The moonlight painting of Romanticism is typified by recurrent genre motifs such as the reflection of the moonlight on the surface of the sea, or backlight, where a dark foreground stands in contrast to an illuminated firmament; but each artist approaches the moonlight motif with a distinctive gaze. In C.W. Eckersberg's almost black *Moonlight Painting* (1821, p. 4) we meet a tamed nature, as calm as a well-tended garden. The circular full moon hangs like a shiny coin in the top corner of the picture. Everything seems drawn with a ruler.

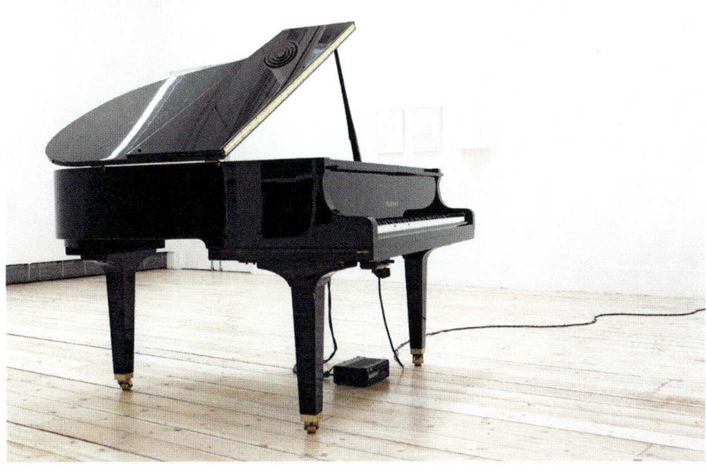

cat. 107
Katie Paterson
Earth – Moon – Earth (Moonlight Sonata Reflected from the Surface of the Moon), 2007

The mast of the boat divides the picture surface into geometrical fields, and the moonbeams are reflected in the neat ripples on the water surface with graphic precision. The night is mild. A different drama is present in J.C. Dahl's painting from the same year, *The Bay of Naples seen from a Grotto* (1821). Inspired by a journey to Italy in 1820-21, Dahl created a number of intriguing moonlight studies that explore the landscape in and around the Bay of Naples, a popular destination of the time for the cultivated and prosperous. In this little painting the moon is half-hidden behind a cloud. We look out at the sky and sea from a grotto, which provides the opportunity to paint a picture in backlight. This way Dahl achieves a special mystique, since all that is closest to us we cannot see. It is about the beyond. In the middle of the picture the volcano Vesuvius appears with its active, lava-dripping crater, as a red-orange counterweight to the grey-white light of the moon. The bowels of the earth and outer space meet – volcanic smoke and clouds mix against the sky, and midway between the darkness of the grotto and the light of the sky a small human being sits fishing. A quotidian task measured against the vastness of the elements. There is a geological awareness in this picture with the grotto – we are so to speak in the earth's interior. The lava, the rocks and the moon draw a line from Dahl all the way to contemporary art's preoccupation with geological processes and the planetary perspective on earth.

British Joseph Wright of Derby travelled, like J.C. Dahl, to the Bay of Naples (1773-75) and in the picture *Virgil's Tomb, by Moonlight* (1782, p. 36) painted a ruin which was thought to be the Roman poet's burial place, and which had become a well-known attraction at the time. In Wright's painting the huge rock tomb sits like a skull in the landscape, and the sharp light of the full moon makes the cloud formations draw a white pattern on the sky. From the inside of the tomb one can make out a faint glow from a small lamp. Wright was the artist of the period who most consistently worked with moonlight, and he has been given the by-name "the painter of light" because besides moonlight he worked in detail with the effect of candlelight, and gradually also incorporated the new lamps of industrialization in his paintings. Wright was interested in the technological developments of his time and was a member of a community of prominent industrialists, artists and scientists known as The Lunar Society. Their meetings were held on evenings with a full moon – for the practical reason that the moonlight made it easier and safer to find their way home after the meeting. This was soon to change, for Wright's time and the subsequent years were characterized by inventions of new artificial lighting types. In 1792 the Scottish engineer William Murdoch invented what is considered the first gas lamp, and in the following decades a succession of new types of gas and electric light developed; then in 1880 Thomas Edison produced a 16-watt light bulb. The intense interest in moonlight that can be seen in the work of the painters of the period thus developed in the very years when the moon gradually, but inexorably, succumbed to competition as the primary light source of the night. It is tempting to conclude that Romanticism's obsession with moonlight was in reality a response to the new technological advances; a result of the sudden new importance of moonlight when other lamps began to light up the night.

Contemporary art stands at the peak of this development of light. In the comprehensive photographic project *Fullmoon* the artist Darren Almond has sought out particularly dark landscapes – mountains, coastline, forests and rivers – which he has photographed by the light of the full moon. By photographing with a long exposure Almond is able to capture so much light that a colour picture emerges from the landscape of the night. Soft, burnt colour shades give access to a new experience of nature – and a special drama arises in the picture because the long exposure of the camera emphasizes the contrast between what moves and what stands still. Stars lash the sky like swaths of rain, and the water moves diffusely as smoke among rigid rocks. In these pictures the camera reveals a new view of the landscape – an experience of nature to which we normally have no access. And with a clear nod to the landscapes of Romanticism Almond turns the moonlight into an object for the contemplation of great nature.

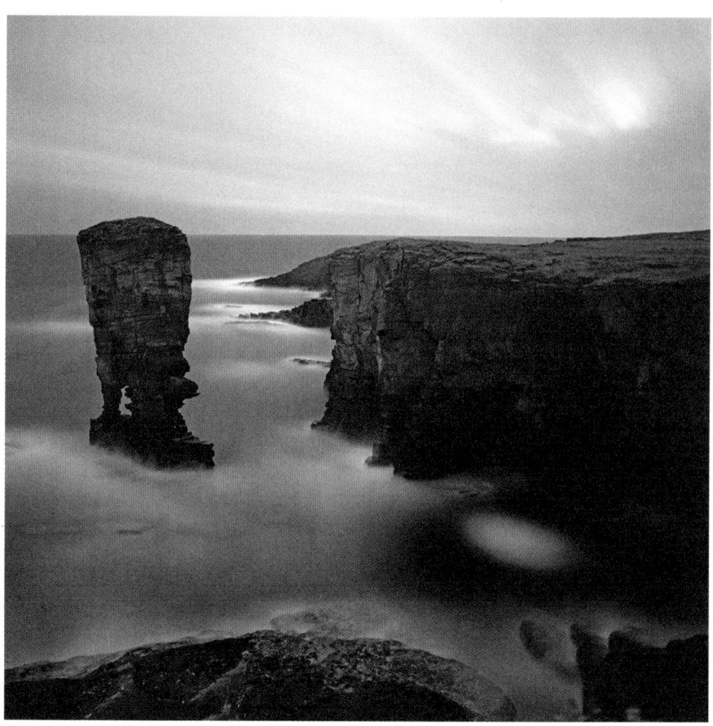

cat. 11
Darren Almond
Fullmoon@Yesnaby, 2007

Observation and Speculation

As a focus of cultural fascination the moon appears at the centre of a conflict between the observing, surveying gaze of the sciences and a spiritual or decidedly mystical endeavour. The French author Victor Hugo tellingly sums up this antithesis when, after a visit to the Paris Observatory in 1830, he writes: "You see the moon, and this figure of the unexpected emerges before you; and you find yourself face to face with this atlas of the Unknown. The effect is terrifying … You grow dizzy at the sight of a world suspended in emptiness."[8] Between the lines in Hugo's statement sounds the echo of an outer space devoid of God, which he has himself seen to be lifeless.

The scientific observation of the moon has a long history, and one of its main figures is Galileo Galilei, who in 1609 constructed a telescope on the basis of what he had learned about this brand new technology. Over several nights in the winter of 1609 Galileo Galilei observed the moon through a telescope from the campanile of San Giorgio Maggiore in Venice. There he sketched the surface of the moon, and the result was the first published pictures of the surface of a world beyond the earth. The drawings, which were converted to watercolours and afterwards published in Galileo's work *Sidereus Nuncius* (1610, p. 72) changed astronomical science forever. Galileo's background in both science and art meant that he could draw on several disciplines in his investigations, and it was his knowledge of shadow formation in painting that made him connect the darker areas of the moon with irregularities in its terrain.[9] As Galileo wrote: "The Moon is by no means endowed with a smooth and polished surface, but is rough and uneven and, just as the face of earth itself, crowded everywhere with vast prominences, deep chasms, and convulsions." And thus he challenged the prevailing idea, put forward by the ancient philosopher Aristotle, that celestial bodies were perfect spheres – ideal in contrast to the flawed earth – and that the moon was round, a translucent ball. Or as Dante later wrote: "Luminous, dense, consolidate and bright/ As adamant on which the sun is striking ... the eternal pearl."[10] Galileo's story is in many ways a story of the battle between religion and science. His assertion of the heliocentric world-picture made Galileo the victim of repeated inquisitions, and because of his discoveries he spent the last long period of his life in house arrest. With Galileo the moon once and for all became an object of scientific study. The precious series of paintings *Astronomical Observations* (1711, pp. 78-79) by the Italian Donato Creti, which has been lent out by the Vatican Museum to Louisiana's exhibition, tells the story of the dependence of early science on religion – for the series was made as a commission from Count Luigi Marsili, who donated it to Pope Clement XI to convince him of the importance of establishing the first public observatory in Bologna. The artwork fulfilled its function as a political argument – the observatory was established – and the pictures remain today as testimony to the insights of the age into the planets of the solar system. Each of the planets known at the time has its own tableau, where they are studied or observed through telescopes by men and women – and in the panel devoted to the moon its craters are neatly executed with clear inspiration from Galileo's drawings.

In the 1800s accurate maps of the moon had won wide popular acceptance in Europe, and with the invention of photography a new picture technology arose which was quickly applied to the moon. The Scottish engineer and artist James Nasmyth created a series of profoundly fascinating pictures of the moon on the basis of studies through a sophisticated telescope. As a technological pioneer and inventor Nasmyth was preoccupied with the new camera technology, and in the wish to combine it with astronomical observations he created a series of profoundly fascinating illusions of the moon landscape – viewed as if from a position standing on its surface. Paradoxically, the realistic pictures arose through several layers of modelling: on the basis of observations Nasmyth modelled detailed reliefs in plaster, which he photographed against a dark background in strong light – so that contours and shadows of the surface of the moon emerged clearly. The pictures were published in the book *The Moon: Considered as a Planet, a World, and a Satellite* (1874, cover and p. 75), which has later influenced the depiction of the moon landscape from film art to the photographs brought back by the Apollo mission.

One of the most ambitious moon models of the time also results in the photograph that perhaps best captures the ambivalent role the moon plays in the modern western mind. In 1849 the astronomer Johann Friedrich Julius Schmidt persuaded Thomas Dickert, the curator of the Natural History Museum in Bonn, to produce an ambitious model of the visible side of the moon, based on a detailed moon map from the 1830s.[11] The almost six-metre-tall model was produced in plaster and reproduced more than 20,000 details from the surface of the moon – craters, 'seas', mountain ranges – and the model was later sent on an exhibition tour of the USA, where it ended up in the Field Columbian Museum in Chicago.[12] In a photograph from there (p. 3) we see the moon model in a historicizing interior with stucco work and carved panels, and beside it a uniformed guard stands keeping watch. As a monument to the cartographic ambition of completeness the enormous moon model reminds us of Jorge Luis Borges' short story "On Exactitude in Science" (1946), which humorously points out the paradox of cartography in the tale of a fictive empire where map-making has reached such an advanced stage that a map of the empire fills – the whole empire. Everything is there. As such the photograph speaks of scientific ambition and precision – and says that the moon, after centuries of intensive observation, has become more tangible for humanity. But at the same time the photograph is quite grotesque. Is the small human being who keeps watch by the huge moon trying to prevent it from slipping away? The patinated black-and-white filter of the photograph filter creates a brief illusion that it *is* in fact the moon that has been fetched down to the earth and into the museum. Under the auspices of science the moon detaches itself from logic as the mythical, mystical image-machine it also is.

The Moon of the Imagination

It may well be that the moon can be mapped, but it is also the territory of the imagination. Well nigh all cultures have myths, gods or rituals associated with the moon, and art and folklore are both rich in notions of how the moon affects the human body and psyche, as described in more detail by E.C. Krupp in this catalogue (p. 40). A strong current in modern art insists on the role of the moon in the imagination, on its capacity to evoke amazement – and this applies not least to the artists associated with Surrealism. Astronomy – and occult astrology – played an important role for a number of the most prominent artists of Surrealism – both in the form of references to the latest discoveries of the time and in the use of the firmament as a space for fantasies and imaginings.[13] At the same time many of the movement's artists were collectors of mythological and ritual objects from non-Western cultures – objects which both fascinate through their association with religious myths and their use in rituals with the object of bringing mankind closer to a cosmic order.[14] For artists such as Joseph Cornell, Max Ernst, Wolfgang Paalen, Salvador Dalí, Remedios Varo and Gertrude Abercrombie the firmament becomes a space of poetry.

Remedios Varo was deeply interested in astrology, and in her paintings human beings and the stars are connected in chains of causality known from the horoscopes of astrology, where personal actions and decisions are associated with the positions of the moon, the stars and the planets. Varo, who for large parts of her life lived in Mexico, was like other Surrealists interested in religious myths, rituals and folk culture, and her paintings involved a wild combination of references to all of this. As can be seen in the little votive image *Icono* (1945, p. 70), where the gilded image of the Christian icon tradition reveals not Jesus or the Virgin Mary – but a mysterious flying building which, with wings, propeller – and one wheel – floats across the starry sky, driven by straps that connect the construction to faraway crescent moons. Several of Varo's paintings underscore how mankind and the firmament are connected. The painting *Portrait of Dr. Ignacio Chàvez* (1957, p. 71) has several of her recurring motifs – crystalline forms, medieval mysticism, constellations – and the connection between stars and humans is visualized by thin lines which like industrial webbing control the slender Gothic-style human figures by means of a kind of celestial mechanics. New doors open – in the world and in people. The connection with the beyond becomes a means of achieving deeper awareness.

In European folk culture the new moon is associated with fertility, while the full moon is both ascribed healing properties and is considered a source of madness – lunacy. Max Ernst's sculpture *Moonmad* (1994 / cast in 1973) stands as a golden fetish with bulging eyes and a wild, open mouth – from its round head two crescent-shaped horns stick out. The sculpture testifies to Ernst's inspiration from the wooden figures of the West African Kota people and looks like a satyr or a diabolical idol. Under the light of the moon the mind opens up, and in

cat. 54
Max Ernst
Moonmad, 1944

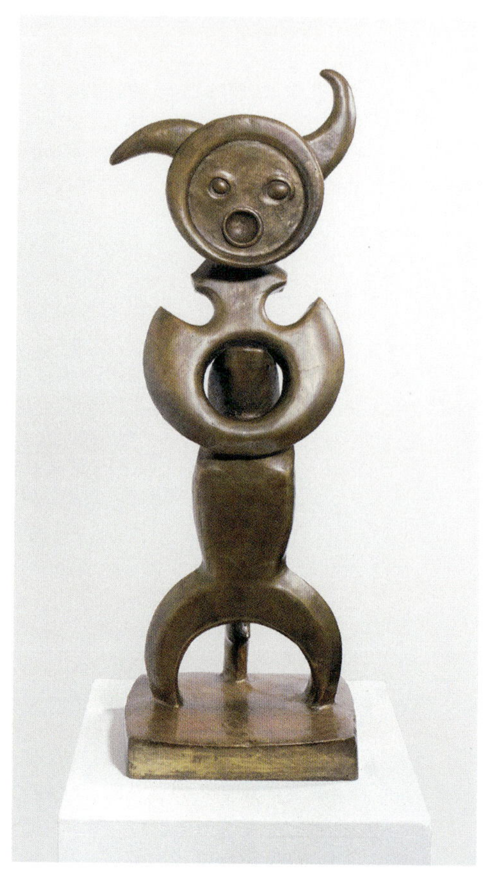

cat. 37
Joseph Cornell
Untitled (Solar Set), c. 1956-58

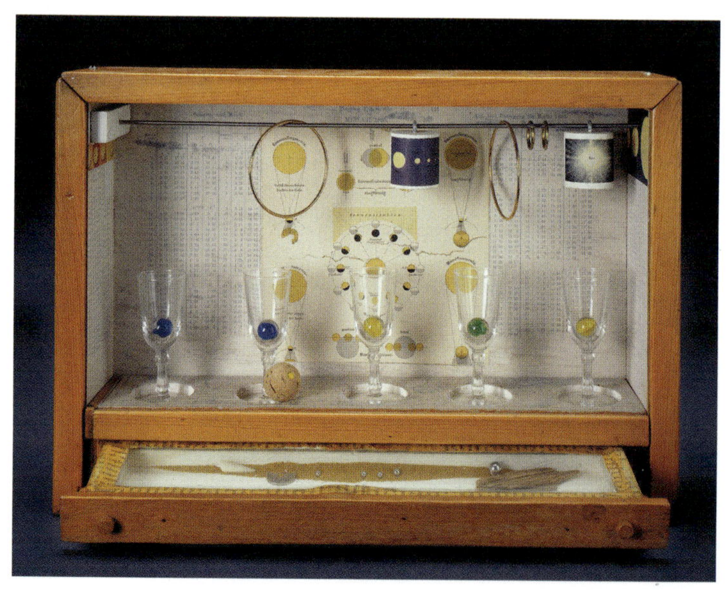

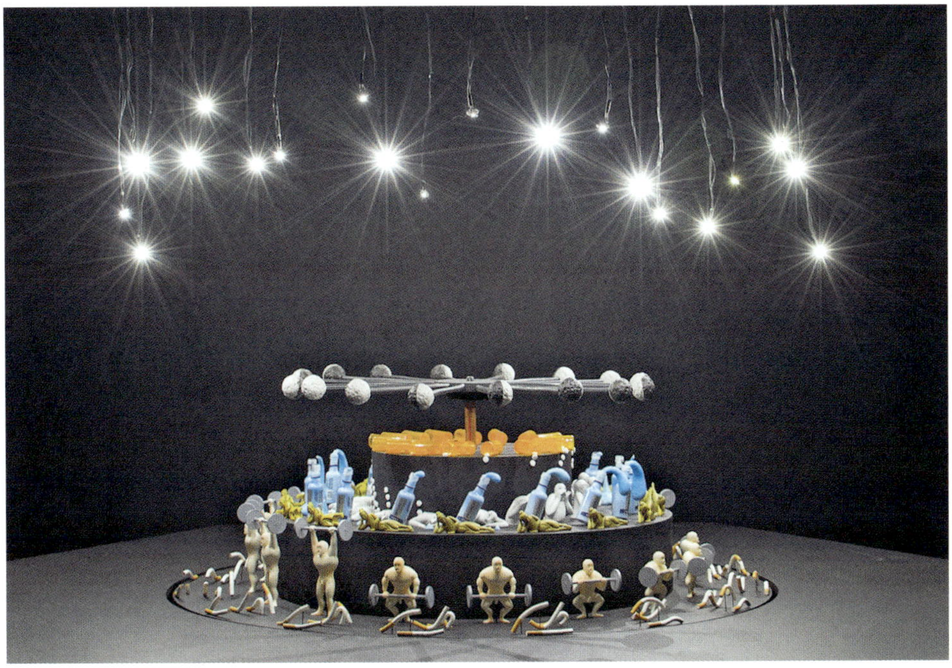

cat. 65
Camille Henrot
October 2015
Horoscope, 2015

Ernst's work it is in the form of a naked primal scream that throws off the mantle of rationality.

A contemporary artwork which also makes the connection between the moon and a state of mind is Camille Henrot's mechanical sculpture *October 2015 Horoscope* (2015), which has the changing cycles of the moon rotating in a continuous circular motion over an array of small figures: with mechanical repetition muscular men lift a lever up and down while female Buddha-like figures seesaw with their legs pilates-style, cigarettes hop along like caterpillars in a cartoon film, and pills from spilled medicine vials rain down over a small human being who alternately smiles hysterically and collapses in total helplessness. A mentality-historical hamster's wheel. The existential absurdity is quite palpable – but in Henrot's work it is the starting point for a social critique. The rotation of the moons becomes the motor for a both humorous and sombre comment on the mental health of our time, which attempts to solve systemic societal problems by taming the body with 'wellness' and medicine. Are we actually ourselves masters of the world we create? And how does it affect us?

Henrot's humour and use of play and games are related to the Surrealist artist who most consistently used cosmic themes: Joseph Cornell. Astronomical themes run as a constant strand through Cornell's works, where the grand scale of the universe is placed with playful ease within our reach. Like other Surrealists, Cornell worked with found materials and unorthodox combinations of objects, but from a very early stage he developed a distinctive visual idiom based on small 'boxes' in which he exhibited selected objects in tableaus, at one and the same time evoking memories of systematic scientific presentations and childhood collections of the hidden treasures of the roadside ditches. In a set of soap bubbles from 1941 a moon map lights up in a box lined with black velvet, flanked on each side by finely carved white pipes and with small pieces of coral in the foreground as allusions to the connection between the moon and the sea. The themes of astronomy are interpreted by Cornell with physical and emotional resources, and the small, intimate boxes set the scene for a personal gaze at the objects which the square framing insists belong together. In the box *Untitled (Solar Set)* (c. 1956-58) an astronomical diagram of the interaction of sun, earth and moon forms the background for an arrangement of liqueur glasses, each with a small glass ball, as if the planets of the solar system have fallen down on the lunch table. Above the glasses hang metal circles as lines of orbit which can be slid forward and back, and beneath it all a sandy beach with shells and driftwood can be pulled out in a drawer. A cosmic marvel whose surreal effect arises in the playful insistence that the whole solar system can easily be folded into the small box. There you have it: the cosmos in a crate.

Colonization of the Moon

A much more tangible version of the moon can be found in the art that deals with space travel. In this the moon is in fact a place you can travel to. A central section of the exhibition is concerned with the aesthetics of the Space Age: the artistic manifestations that came in the wake of the Sputnik's orbit in 1957 and the subsequent space race; a period that Stephen Petersen describes in detail in this catalogue (p. 84). In the years around the first manned moon landing with Apollo 11, in 1969, the race for the moon was an all-dominating cultural agenda whose philosophical implications preoccupied the major artists of the period – from Robert Rauschenberg through Kiki Kogelnik to Yves Klein, among many others – with equal proportions of optimism about progress and ideological criticism. Today, fifty years after mankind's first steps on the moon, we find ourselves in the midst of a new space age. And in this case the moon is not just a destination, but a springboard: from being man's remotest destination, moonbases are now to be the earth's outposts for travel out into the universe. The architectural firm Foster+Partners has developed the design for a 3D printed moonbase for the European Space Agency ESA – a building that is to be printed on the moon from available materials and offers great promise of production in space (p. 119). Factories on the moon? Hotels? What world can we imagine out there? The artist Rick Guidice was engaged in the 1970s by NASA to visualize possible space colonies. His picture of the gigantic wheel-shaped Torus colony for 100,000 inhabitants is organized in a surprisingly traditional pattern as residential neighbourhoods. Suburban life on a totalitarian scale is offered with the universe as a backdrop. So much for the endless possibilities for utopia.

The efforts of art to put us in contact with the grand scale of the universe have a new topicality today, now that humanity has made contact with the deep time of geology. In recent years researchers have been using the term 'the Anthropocene' to indicate that the earth has entered a new geological age in which mankind affects the earth at a planetary level – for example by moving more mass than other factors such as wind conditions, erosion and plate tectonics.[15] One of the theorists who have worked most consistently with the Anthropocene is the French philosopher Bruno Latour. In the book *Facing Gaia* (2017) Latour describes a new world order in which the relations among science, politics and culture must be rethought. With reference to Romanticism's view of nature, Latour describes a new situation where the opposition of the eternity of nature to ephemeral mankind no longer makes sense. For the cosmos itself is in motion. Glaciers melt faster, the oceans are rising and species are disappearing at a higher rate than the development of the political and cultural processes which should prevent this happening. Today, he writes, it is easier to imagine the end of the world than the end of capitalism.[16] The destructive power of man is more stable than any force of nature.

Latour's point makes an impression at a time when a new space age is under way, driven by both nation-states and commercial enterprises, which aim at the colonization of the moon with a view to mining in outer space. And several of the contemporary artworks in the exhibition in fact deal with mankind's imprint on the universe. In *Project Adrift* (2016) the documentarist Cath Le Couteur investigates the world of space garbage, which for many people is an unknown aspect of the close orbit around the earth. As a result of space travel, since the 1950s, more than 100 million pieces of man-made space garbage circle the earth: decommissioned satellites, objects such as cameras, toothbrushes, thermal blankets and tools lost or thrown away by astronauts, rocket parts and fuel tanks, which are smashed to pieces in collisions and become a multitude of objects.

The night is full of more than twinkling stars, as the artist Trevor Paglen underscores with his photographic series *The Other Night Sky* (2010-11, p. 117), which shows how the underlying militaristic agendas that have followed the space programmes since the beginning are still active today. Paglen furnishes his series of sublime shots of the sea of stars in the night sky with sobering titles which point to the secret military surveillance satellites that constantly orbit the earth. The artist Hito Steyerl's docufiction video work *ExtraSpaceCraft* (2016) is set among the ruins of an observatory north of Iraq, where a militarized space programme based on surveillance by drones monitors the sky over Kurdistan. The spirit of the age is not about looking curiously out into space, it is suggested in the video's edited-together mix of video game aesthetics and war – for we go into space to monitor the earth. The current plans to establish moonbases as starting points for missions in outer space raises many issues, as a new cosmic territorial race arouses memories of the colonial period's hunt for land and resources. Who owns the moon? Two years after the Soviet Union succeeded in putting the Sputnik into orbit around the earth, the UN established The Committee on the Peaceful Uses of Outer Space (1959), with the particular aim of preventing the placing of nuclear weapons on the moon and in space as such. The Committee established a space treaty in 1967 which lays down the

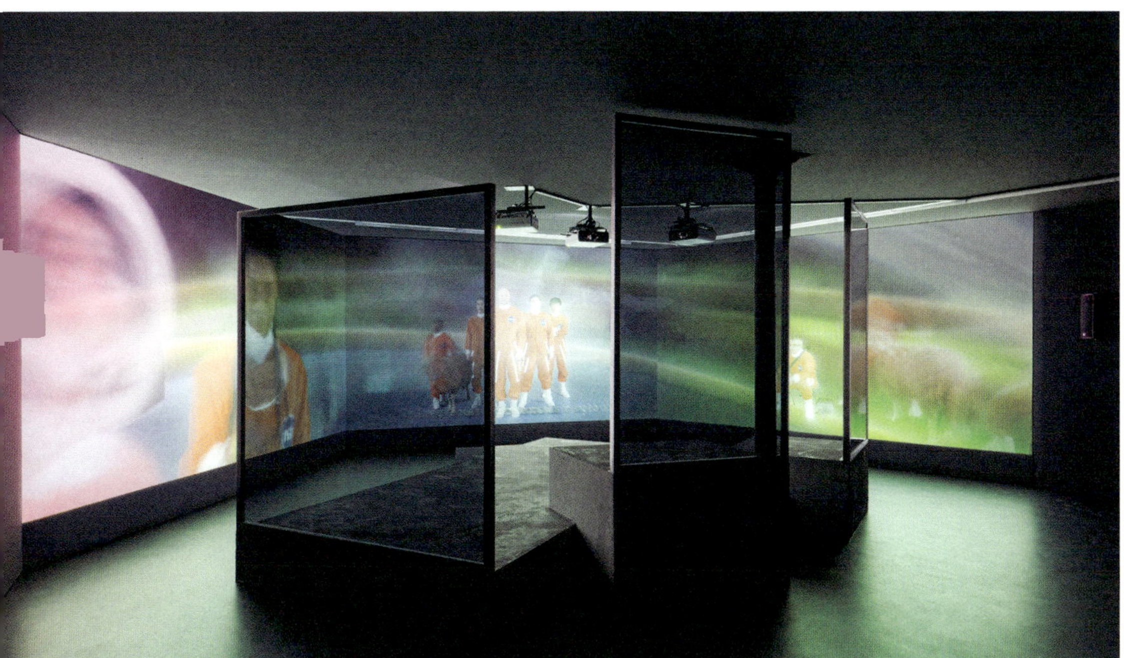

cat. 163
Hito Steyerl
ExtraSpaceCraft, 2016
Installation shot,
Basel Kunstmuseum

framework for legislation on space, and states that outer space is open for peaceful exploration by all nations, while "outer space, including the moon and other celestial bodies, is not subject to national appropriation by claim of sovereignty, by means of use or occupation, or by any other means."[17] The moon was thus defined as the common property of mankind, part of "the global commons" in line with the sea bed and the atmosphere. Nevertheless, plans to send robots to the moon to extract minerals, establish factories and post researchers and astronauts to a permanent moonbase seem more within reach than ever. The current plans for space colonization tellingly reflect the balance of power on earth – with great powers like the USA, China, Russia and Europe as active players. Large parts of the population of the world will not see themselves represented in this world of the beyond, but must continue to stare with longing from the earth at the moon – as mankind has always done.

As I wrote at the beginning of this article, our pictures of the moon can tell us something fundamental about who we are. The moon reflects the thoughts about existence and about the place of human beings in the world that are typical of the time. What is important to us, to what do we ascribe value, how do we create meaning in the world? Ask the moon.

1 Armin Zweite, *Hiroshi Sugimoto – Revolution*, 2012
2 Christine Kintisch, *Katie Paterson – Inside this Desert*, 2016.
3 Sabine Rewald (ed.), *Caspar David Friedrich – Moonwatchers*, 2001.
4 Ibid.
5 Arthur Schopenhauer, *The World as Will and Presentation*, Volume Two, Routledge, 2016, pp. 424-425.
6 Max Weber, *The Sociology of Religion*, 1971, p. 270.
7 The original title of the sonata is *Piano Sonata No. 14, Op. 27, No. 2.* Since Rellstab's remark the title "Moonlight Sonata" (German "Mondscheinsonate") has been used in English and German publications, and it has been universally known ever since under this title.
8 Jean Clair, "From Humboldt to Hubble," in Jean Clair (ed.), *Cosmos – from Romanticism to the Avant-garde*, 1999, p. 20
9 Erwin Panofsky, "Galileo as a Critic of the Arts: Aesthetic Attitude and Scientific Thought," in *Isis*, vol. 47, no. 1, March 1956, pp. 3-15.
10 Dante: "Paradise," in *The Divine Comedy*, translated by Henry Wadsworth Longfellow.
11 Andreas Christoph, "On the 'De-Measuring' of Time and Space in the Models of the World of the Eighteenth and Nineteenth Centuries," in Mirela Altic, Imre Josef Demhardt, Soetkin Vervust (eds.), *Dissemination of Cartographic Knowledge: 6th International Symposium of the ICA Commission on the History of Cartography*, 2016, p. 273.
12 The model, which consists of 116 individual parts, is today in the collection of the Science Museum in London.
13 This perspective is a relatively new field of research. See for example Ashley Lynn Busby, *Picturing the Cosmos: Surrealism, Astronomy, Astrology, and the Tarot, 1920s-1940s*. PhD dissertation from the University of Texas at Austin, 2013 (unpublished), and Kirsten Hoving, *Joseph Cornell and Astronomy – A Case for the Stars*, 2009.
14 Jennifer Field (ed.), *Moon Dancers: Yup'ik Masks and the Surrealists*, 2018.
15 Critics of the concept have pointed out that it is not humanity as such, but rather a specific capitalist societal model, that has given rise to this situation.
16 He has this formulation from Fredric Jameson's publication *Future Cities*, 2003. See Bruno Latour, *Facing Gaia – Eight Lectures on the New Climatic Regime*, 2017, p. 108.
17 https://en.wikisource.org/wiki/Outer_Space_Treaty_of_1967#Article_VII

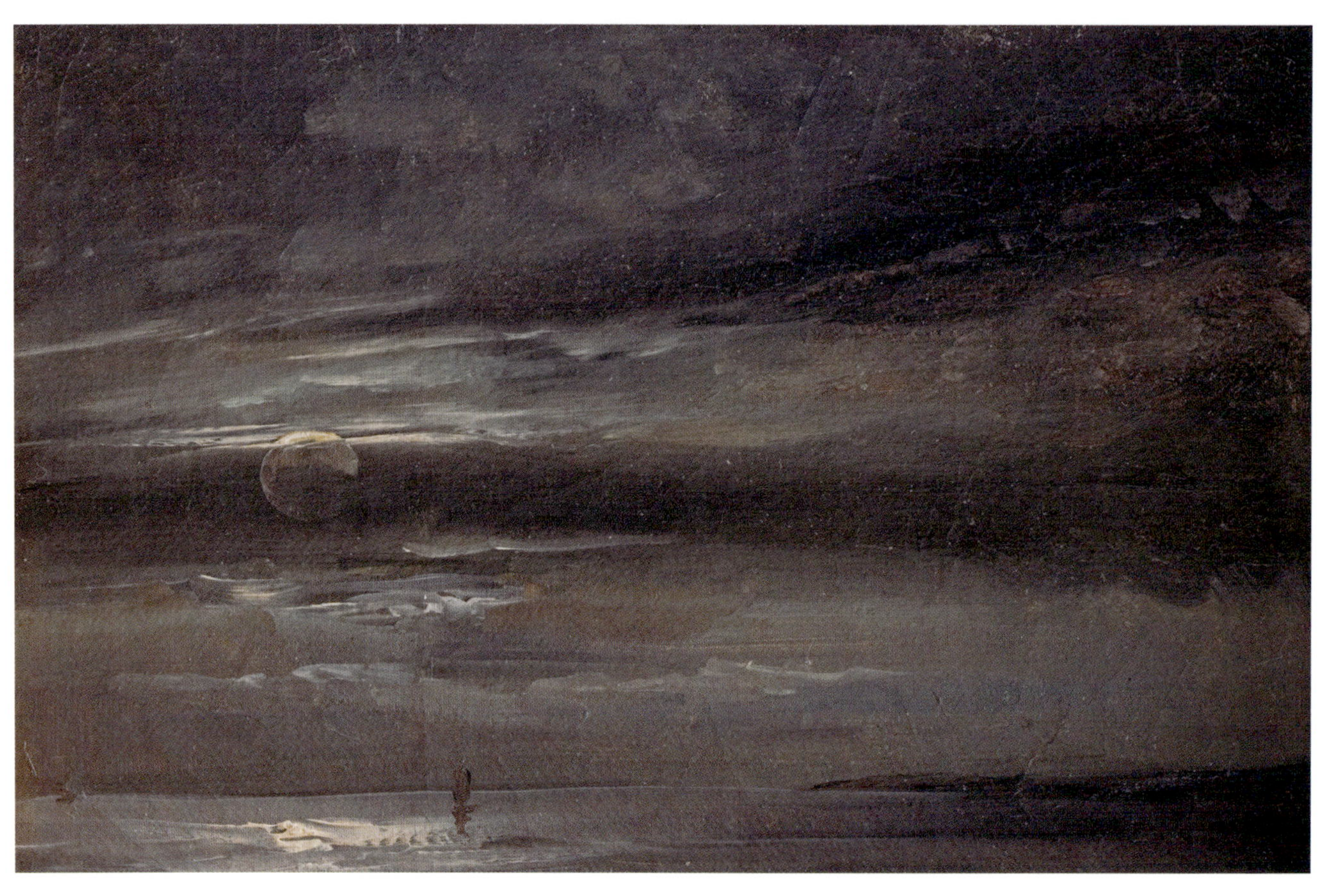

cat. 43
J.C. Dahl
Måneskin over havet, c. 1820-35
Moonlight over the Sea

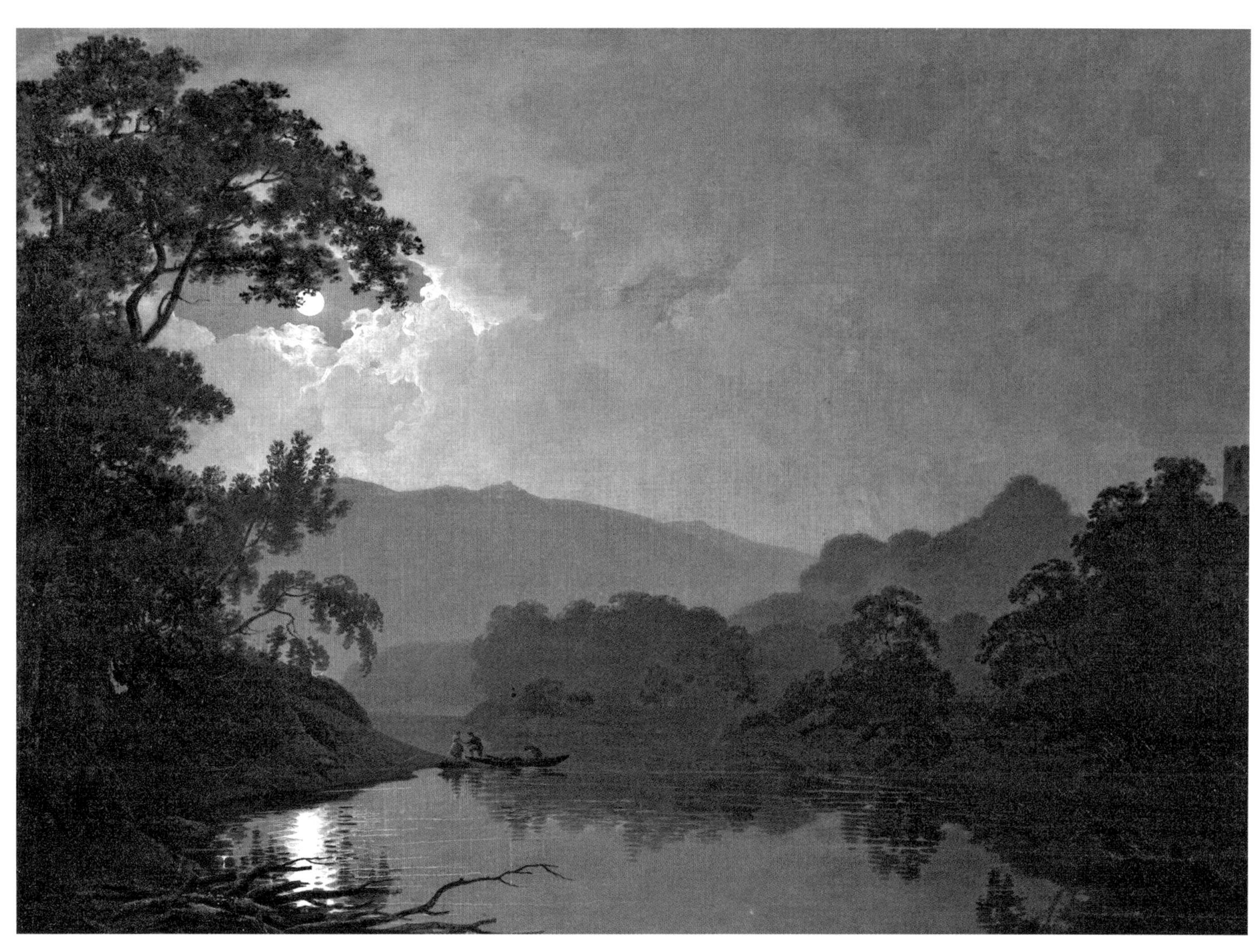

cat. 214
Joseph Wright of Derby
Snowdon by Moonlight, year unknown

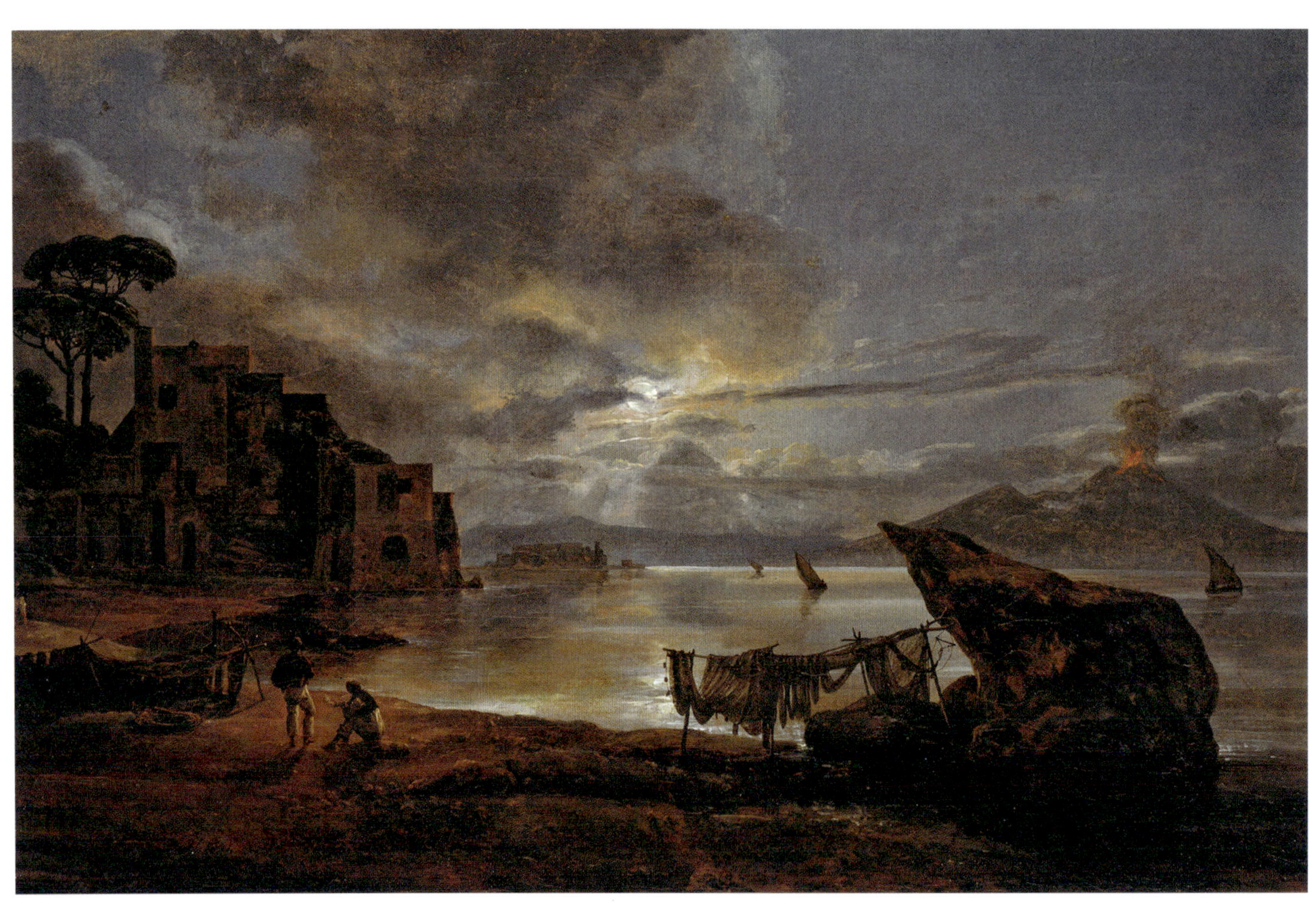

cat. 41
J.C. Dahl
Bugten ved Napoli, 1821
The Bay at Naples

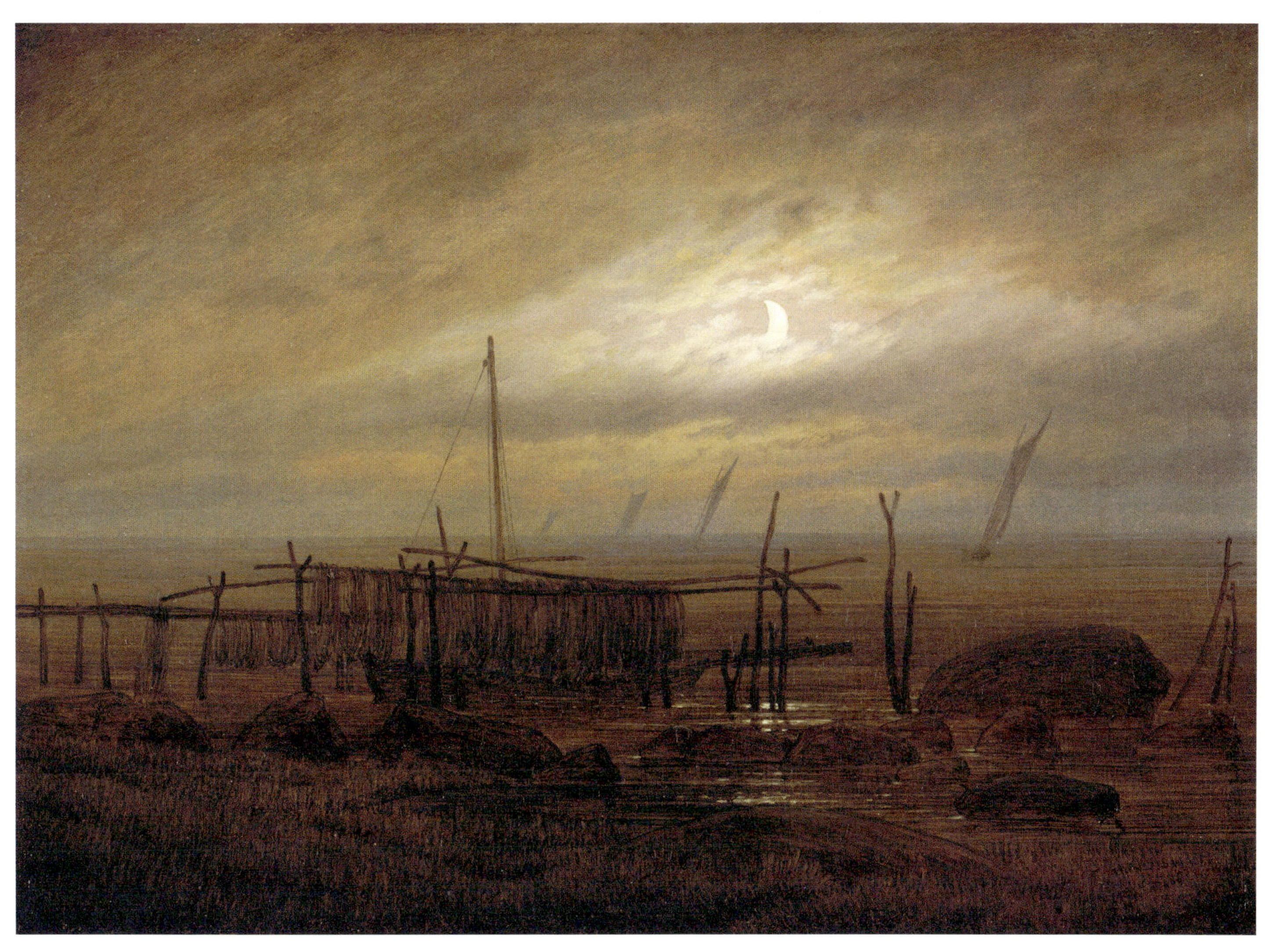

cat. 58
Caspar David Friedrich
Meeresküste im Mondlicht, 1818
Seashore by Moonlight

cat. 97
James Arthur O'Connor
The Poachers, 1835

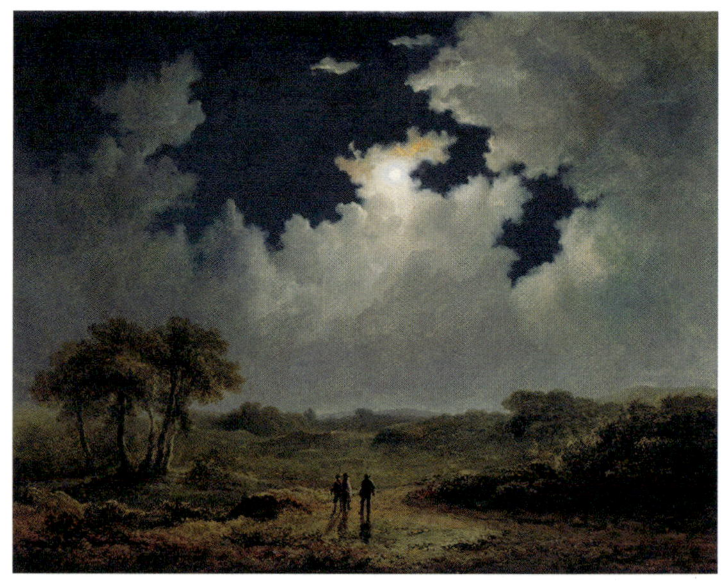

cat. 212
Joseph Wright of Derby
Virgil's Tomb, by Moonlight, 1782

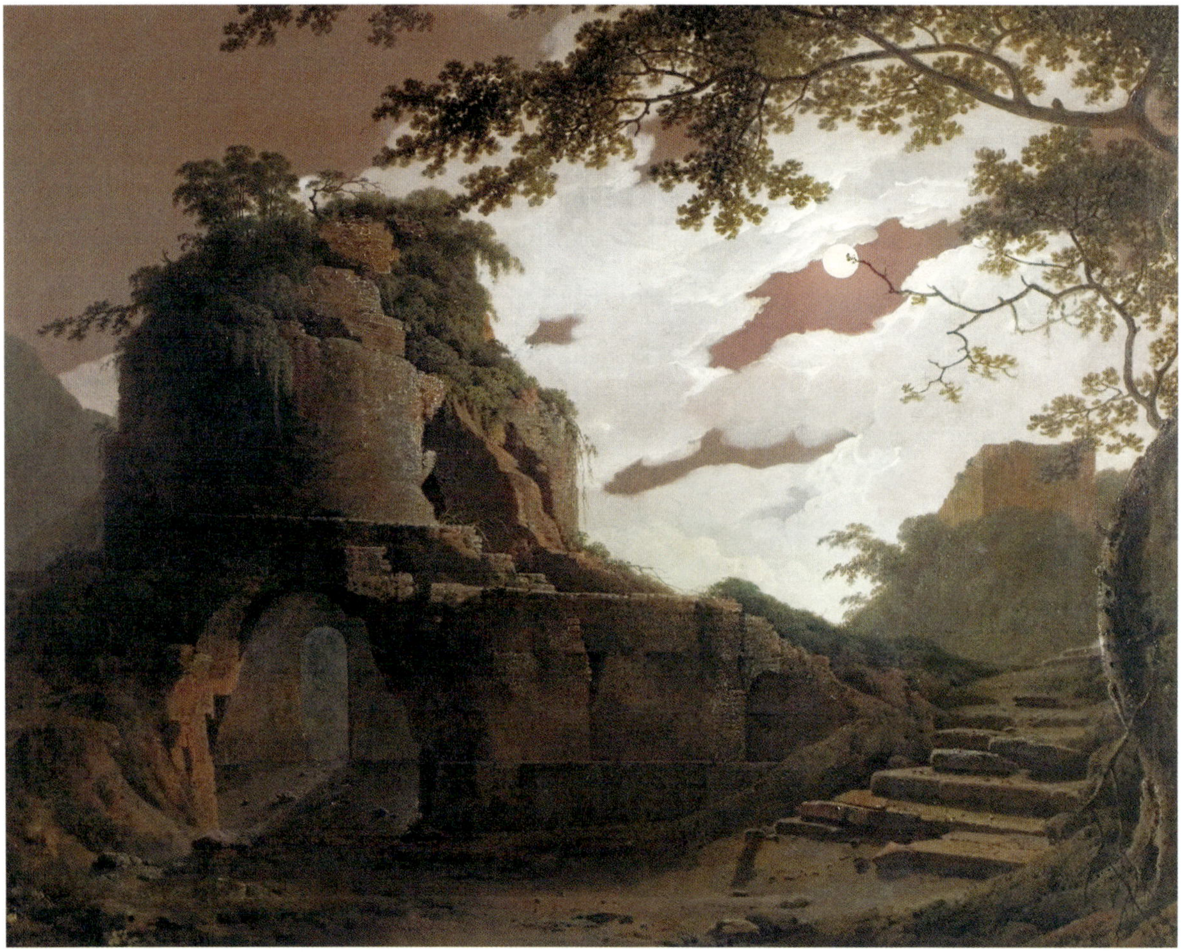

cat. 213
Joseph Wright of Derby
Lake with Castle on a Hill, 1787

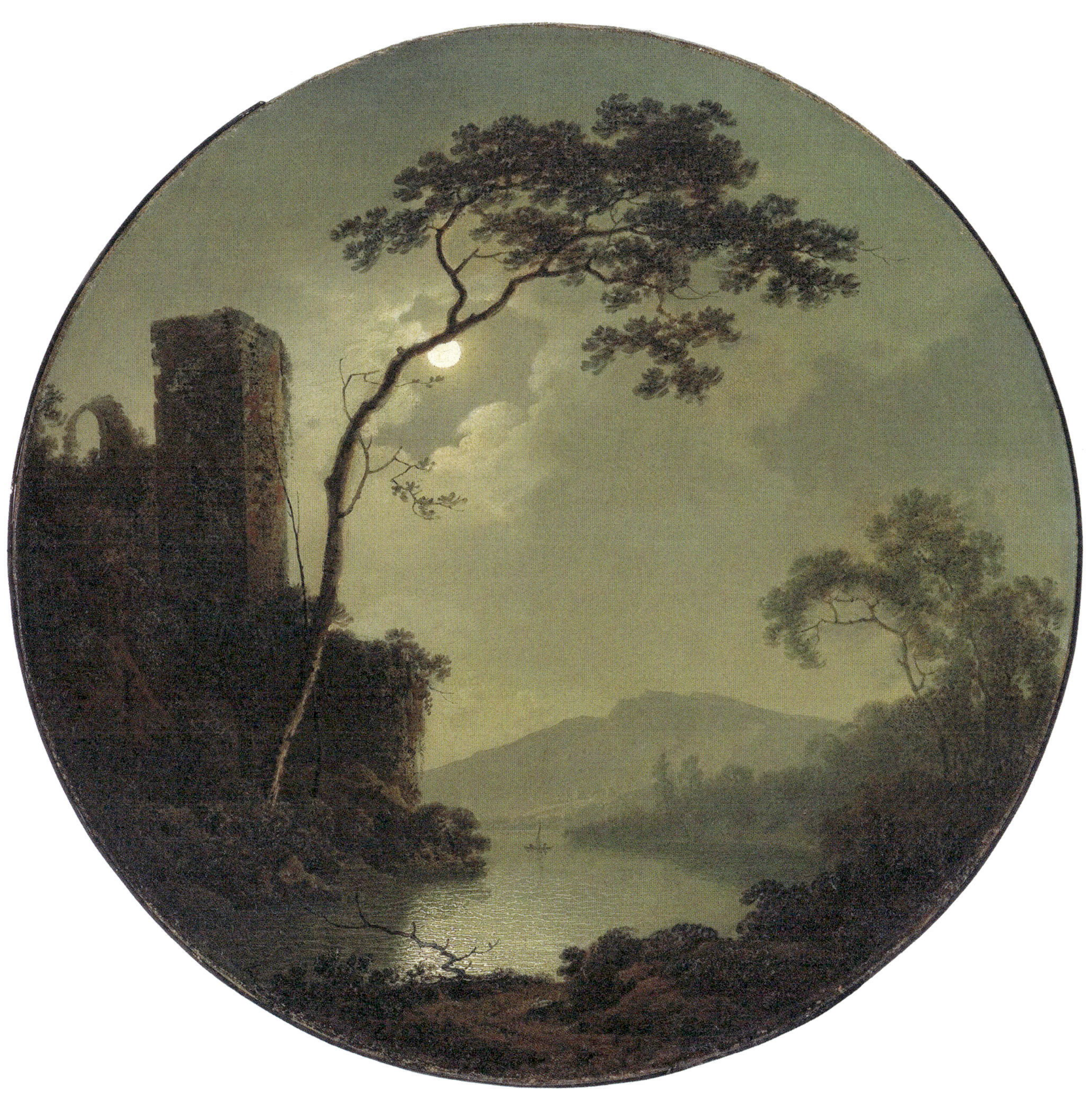

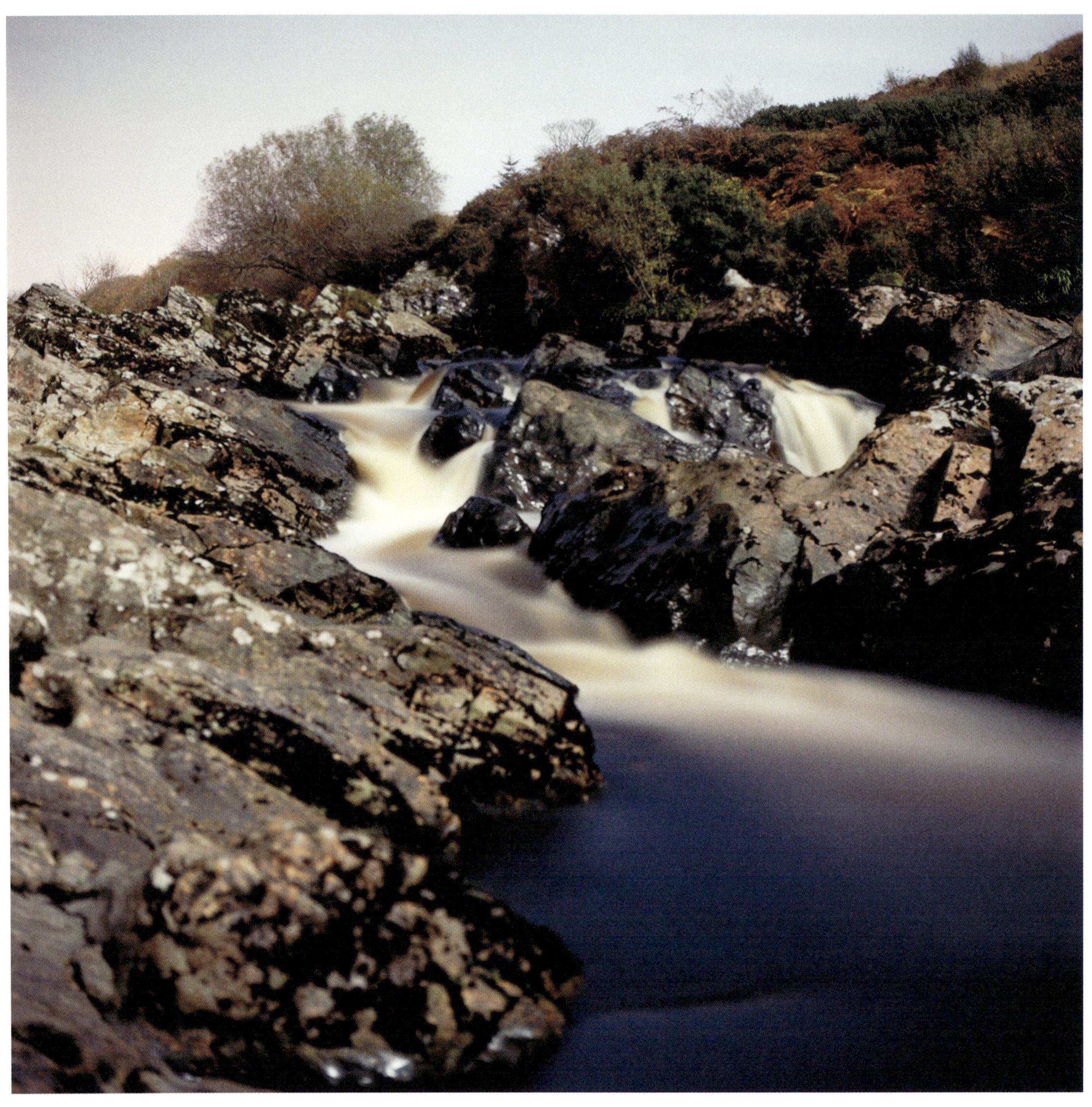

cat. 9
Darren Almond
Fullmoon@Peat Brook, 2007

cat. 8
Darren Almond
Fullmoon@Gordale Scar, 2005

cat. 10
Darren Almond
Fullmoon@Wester Ross, 2007

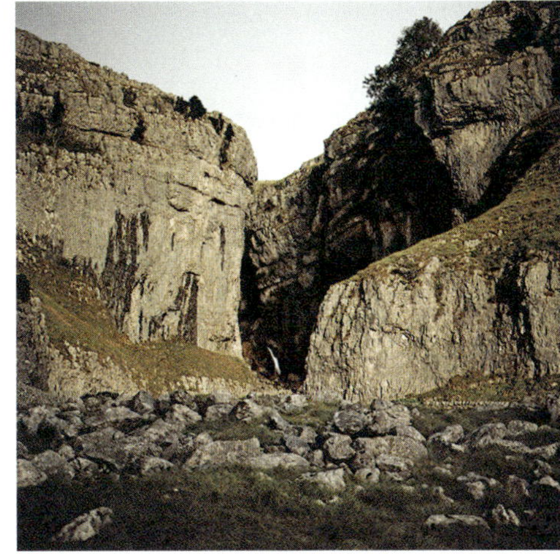

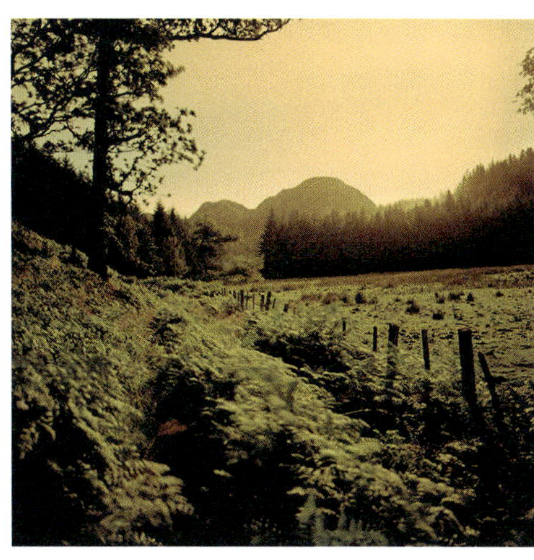

cat. 7
Darren Almond
Fullmoon@Fell, 2005

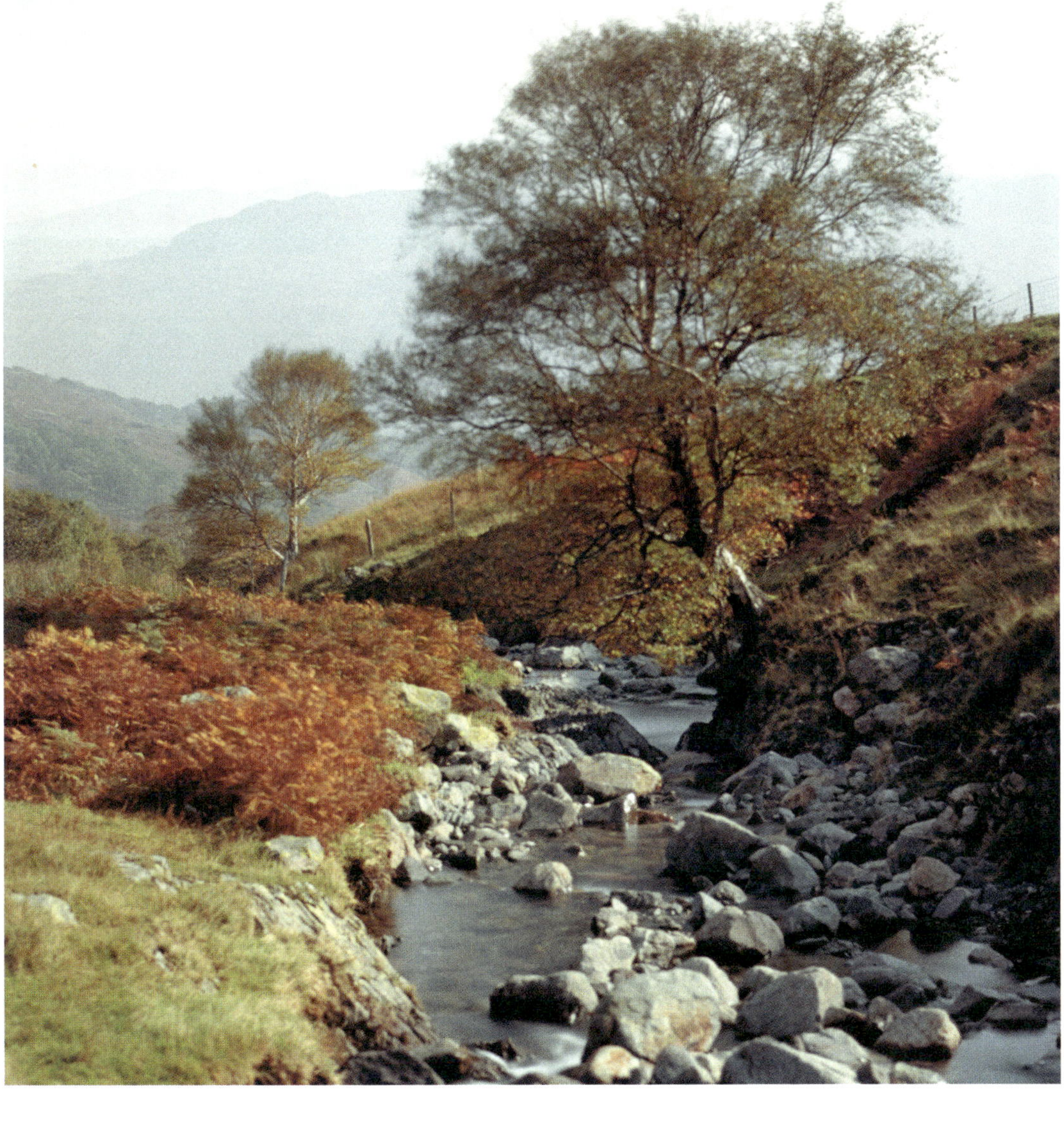

Moon in the night, ruler of the stars,
who distinguishes seasons,
months, and years:
He comes, ever-living,
rising and setting.

Ancient Egyptian text

Lunar Influence

By E.C. Krupp

E.C. Krupp is an American astronomer, writer and expert in the field of archeoastronomy, the study of ancient cultures' perceptions of the heavens. He has been Director of the Griffith Observatory in Los Angeles for more than 40 years.

The moon has been showing up in art and folk culture for millennia. The ancient Egyptian text quoted on this page says it all: The moon is a fundamental element of the natural environment, and we likely have been watching it for as long as we've watched anything. The moon always fulfilled some of the same symbolic functions and carried some of the same meaning throughout the world and across differences in cultural complexity. Even in the Space Age, the moon retains its ability to charm us and accessorize culture, and it is easy to understand why. Despite the light-saturated skies of our cities, the moon is conspicuous. It is the second brightest object in the sky, and it is ever on the move – in its appearance, in its place, and in the time when it is seen. It is born as the new crescent in early evening's western sky. It grows in brightness and form until it waxes into a full disk, when, at full strength, it stays up all night. Then rising later each night, it declines, dies, and disappears, until it is reborn again from the dark of the moon as time midwifes another month of birth, growth, death, and rebirth.

Its phases conveniently bundled the days into packets of time with seasonal value. Its light, its celestial character, its independent movement, and its orderly cyclical change indicated to the ancients it is immortal and

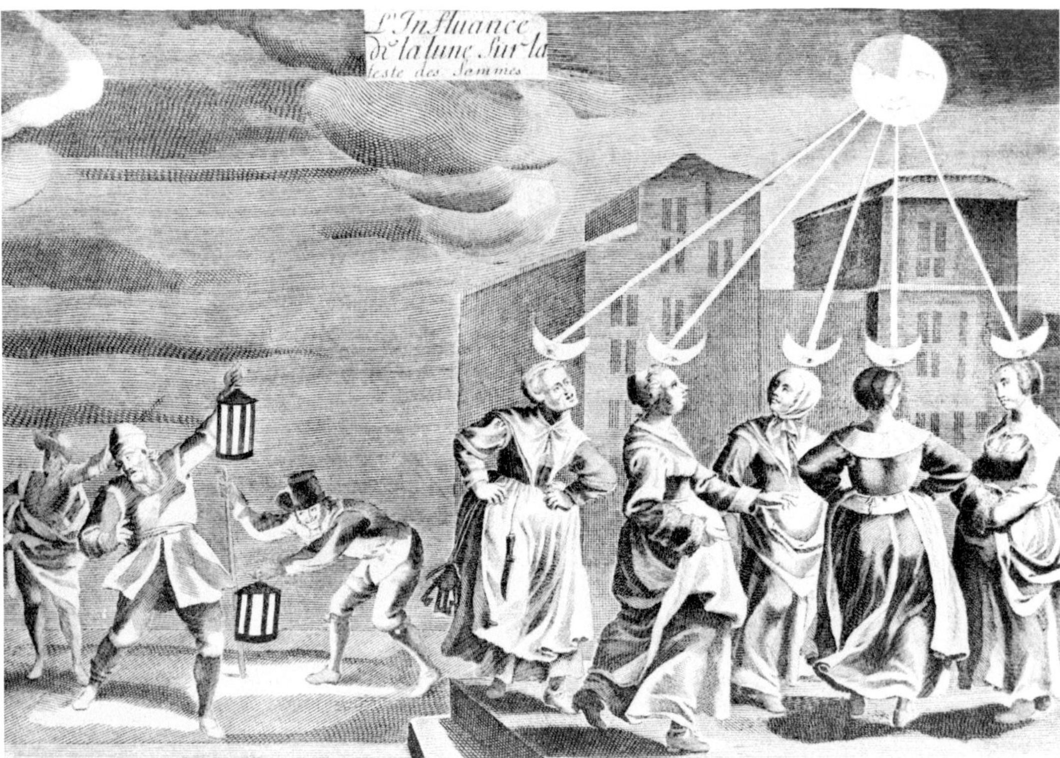

Moon-struck women dancing
French engraving from the 17th century

divine and possesses power over the animals, the plants, people, and the sea. With the powerful illusion of cyclical renewal, the moon seemed to make the world and life what it is. The moon's cyclical pattern of birth, growth, death, and rebirth is recapitulated in the trajectory of our lives, in the lives of all living creatures, in the seasonal growth of plants, in the yearly migration of the sun, and in the annual comings and goings of the stars.

Birth, death, sacrifice, rebirth, immortality, women, fertility, water, the growth of vegetation, werewolves and mental illness – all of these themes meet in the moon because it is the celestial model for, and emblem of, all cyclical renewal. The moon isn't the only agent of celestial order, but it is a headliner on the cosmic stage.

Counting on the Moon

Graphic depictions of the moon may be as old as the neolithic, when our societies of stone-age hunters and gatherers were transformed by agriculture and herding. Neolithic ceramics, six to seven thousand years old, were ornamented with crescents by early farmers in eastern Europe, and neolithic Yangshao farmers at Panshan, in China's Gansu Province, did the same about 2500 BC. No inscriptions verify that these represent the moon, but the crescent is a distinctive and uncommon shape, most often displayed by the moon.

Crescents were also carved into monumental kerbstones that ring the mound of Knowth, a megalithic passage grave in eastern Ireland. Constructed late in the fourth millennium BC by neolithic farmers, Knowth likely operated as a focus of prehistoric ritual. The 22 crescents and seven circles on one of its kerbstones, K52, could represent one cycle of the moon's phases – a lunation. They total 29, and it takes the moon 29½ days to go from one full moon to the next. That interval of time, of course, prompts us to count a month, a term rooted in the word *moon*. The etymology of *moon* originates from the Indo-European word for "measure." By name and by use, the moon was affiliated with measurement of time.

Our ancestors tallied the days that were so conveniently bundled by the phases of the moon. Seasons, however, matter more than the moon, for resources and needs vary seasonally. Seasons are governed by the earth's annual orbit around the sun, which takes approximately 365¼ days. Seasonal sophistication did not require understanding the solar system's geometry, just approximate knowledge of the length of the year, and while the total count of days was sometimes of interest to the ancients, it was far more common to track the passages of the moon to know where in the seasons we were.

The moon completes 12 cycles of phases in 354 days, which is close to the length of the solar year. Many traditional peoples attentively followed the passing months and sometimes named them for seasonal phenomena. The moon doesn't really drive the seasons, but to the ancients the seasons seemed coordinated with the moon. That's an illusion perpetrated by the moon's calendrical convenience. Certain numbers – 12, 13, 27, 29, and 30 – were, then, often associated with the moon, and even though we don't have the owner's manual for Knowth, the 29 crescents and circles suggest the design on the stone was intended to reference the moon – not as a calendar or a record but as an emblem that relied on the moon's behavior and distinctive appearance to make its meaning known.

The moon's costume changes are repetitive and systematic, but they are an illusion. The moon is a sphere and doesn't really change shape. It just appears to do so. The moon's phases occur because the moon is physically round, because it reflects sunlight like a mirror and emits no light of its own, and because the moon orbits the earth. Half the moon is always lit, but the amount of sunlit moon we see depends on where it is with respect to the earth and the sun. When the moon is in the same direction as the sun, the side that faces us is dark, and we see no moon at all. Opposite the sun, the moon is fully lit, and we see it rise, as a complete disk, as the sun sets. As a cold, spherical world in space, the moon sheds only reflected light from the sun, and the cyclical change of the sunlight we see bounced our way turns the moon into a tale of growth and decline.

Symbolizing the Moon

On ancient Crete, about 3,500 years ago, the Minoans placed crescents with people in sacred settings on carved gems to signify the divine power of the moon. At the same time, the bronze-age Mycenaeans on mainland Greece made similar talismans. Some Mesopotamian cylinder seals, engraved about four thousand years ago for authenticating documents, portrayed the moon god Sin in a crescent boat on his celestial commute, and a lunar crescent accompanied some scenes of royal judgment and ritual performance to indicate the powers and prerogatives of the moon. Kassite Dynasty (1600-1150 BC) boundary stones of similar vintage certified land transactions with the rayed disk for the sun, an eight-pointed star for Venus, and a crescent for the moon.

Neolithic and bronze-age graphic allusions to the moon are the oldest recognizable moon symbols, but graphic art is far older than six or seven millennia. Upper paleolithic art – cave painting, petroglyphs, and small carved objects – is attributed to modern humans and can be as old as 40,000 years. Even older pictographs – made at least 65,000 years ago – are assigned to Neandertal people in Spain. Crescent moons and other celestial objects are absent from paleolithic art. Whatever the moon meant to these people, they didn't feel compelled to draw it. The moon's systematic change, however, could be represented symbolically in a numerative composition that concretely leverages the passing of time. Suggestive numbers and cyclical time could mean our paleolithic relatives were more interested in expressing what the moon does than how it looks.

Recycling the Moon

We don't know exactly how the moon eventually persuaded neolithic builders to draw symbols on their monuments, or how the moon in those monument-building chiefdoms became the moon gods of literate civilizations. By the time people committed their sensibilities to writing, however, in bronze-age Mesopotamia, bronze-age Egypt, and bronze-age China, the moon's role in calendars and myth made the moon's most important attribute clear: the moon changes – quickly, singularly, and rhythmically – and is seen as an agent of the cycle of life.

The moon was ancient Egypt's primary sign of cyclical renewal, and Egypt made the ibis-headed god Thoth the calendrical agent of the moon. Affiliated with hieroglyphic writing, Thoth was the sponsor of scribes. In the court of the dead, he recorded the weight of the heart of the deceased against a feather. Sometimes depicted with a crescent cradling a disk, he was in that way crowned with the moon and known as the "reckoner of time" in Ptolemaic-era inscriptions at Dendera. As a timekeeper, he composed the schedule and route for the sun's daily cruise through the sky.

Osiris, as the judge of the dead, was the "Lord of Everything" because he personified cyclical renewal, and that, of course, allied him with the moon. He is graphically linked to the monthly cycle of the phases of the moon in reliefs in the northern half of the ceiling of the main columned hall of the Temple of Hathor at Dendera. In one panel, the 14 days of the waxing moon are depicted as 14 gods, each on a step of a rising staircase that ascends to the full moon, shown as a disk bearing the eye of the god Horus and nested in the crescent of a support stand. Thoth stands to the right, in adoration of the full moon. The hieroglyphic text explains, "the moon-eye is greeted by the sun-eye" on the fifteenth day. That is what happens at full moon: The full moon is opposite the sun, and as the sun sets, the full moon rises.

The adjacent panel to the right on the Dendera ceiling portrays the 14 days of the waning moon as 14 gods seated with the celestial eye in the disk of the moon, which is transported by a boat. Thoth stands in the rear of the boat. On the far left, an introductory panel shows Osiris seated in a celestial boat with the goddesses Isis and Nepthys, and the text unambiguously confirms that Osiris has stepped into the full moon and is the moon.

The Egyptians also personified the moon as the god Khonsu and showed him with a crescent crown. Similar themes framed the divine character of the moon in ancient Mesopotamia, where the Sumerians called him Nanna, and the Babylonians knew him as Sin. His power over ordered change and cyclical renewal made him a judge of the dead in the underworld. He was a light in the darkness, and his phases made him the "great lord" who made the months and completed the years. His reproductive power ensured the harvest and enlarged the herds.

The ancient Hindu moon god is Chandra, whose name means "luminous." He calendars rituals, protects migrating souls, ensures the growth of vegetation, governs the tides, and prompts the rain.

Feminizing the Moon

Egypt, Mesopotamia, and India made a male god out of the moon, but in ancient China, the moon was the home of a goddess, Heng O.

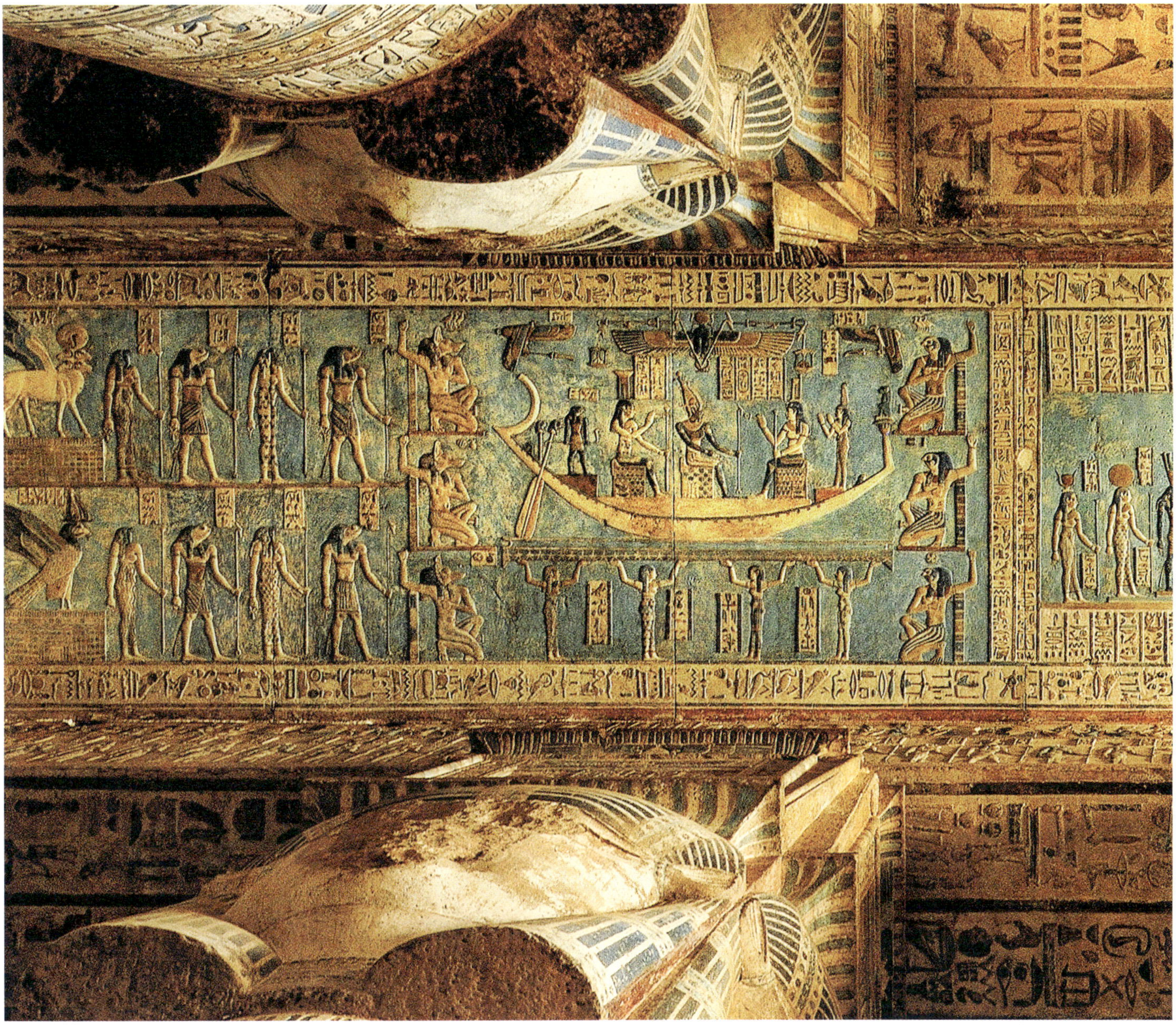

Left panel of the ceiling of the Temple of Hathor at Dendera, Egypt

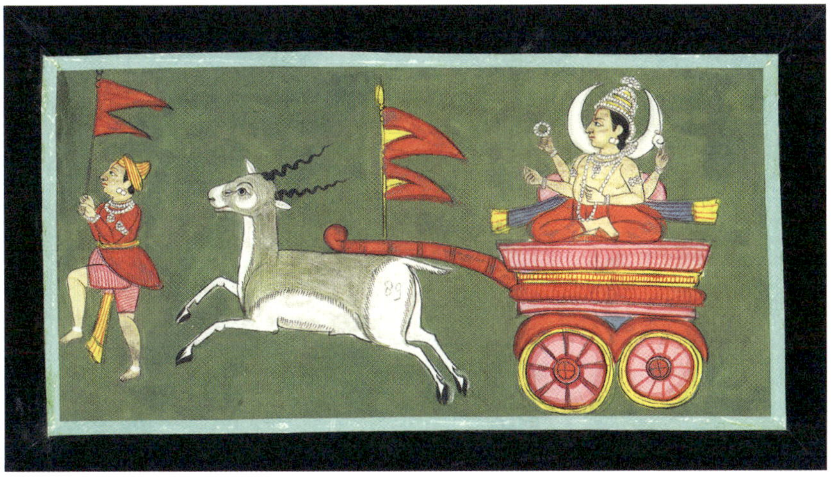

The ancient Hindu moon god Chandra
Gouache on paper

Frans Schwartz
Selene and Endymion, 1865-1917
SMK, National Gallery of Denmark

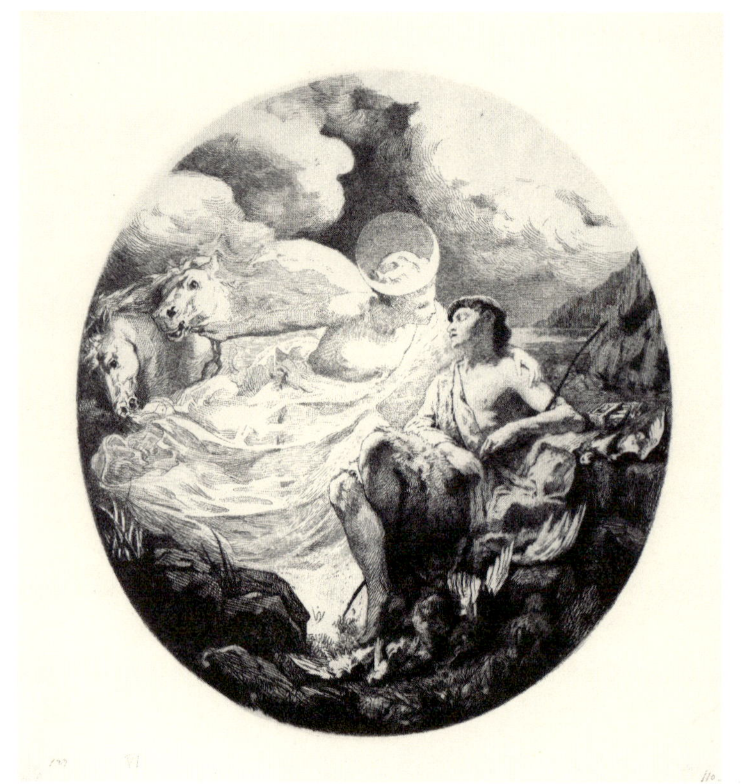

cat. 109
Victor Prouvé
Amour de lune, 1884
Love for the Moon

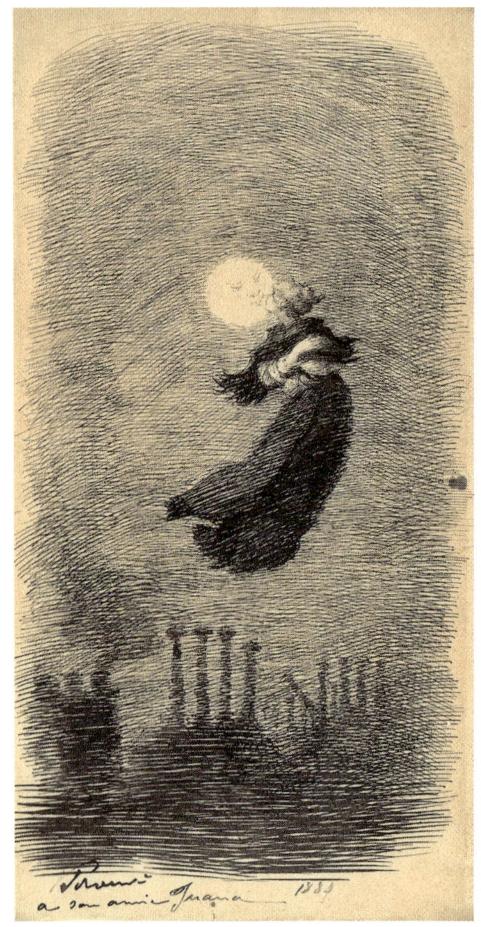

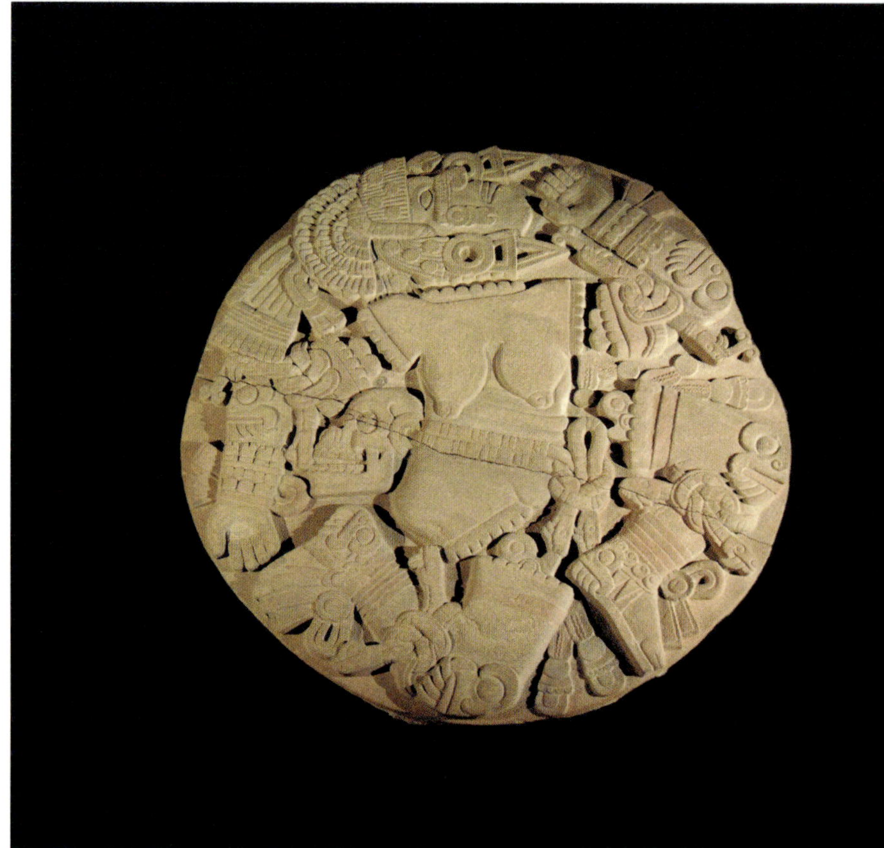

The Aztec goddess
Coyolxauhqui
Replica of monolith

Her unauthorized consumption of the pill of immortality, given to her husband by the Queen Mother of the West, conferred on her everlasting life, but fearful of the consequences for her theft, she fled to the moon. Her immortality reflects the moon's inevitable monthly return, and the Chinese acknowledge her each year with the consumption of mooncakes at the Mid-autumn Moon Festival.

During the Han Dynasty, about two thousand years ago, Chinese artists illustrated the moon in tombs and on funerary banners as a white disk occupied by the rabbit and the toad the Chinese said they saw in the moon, and similar representations of the moon appear in Tang Dynasty royal tombs of the seventh century AD. Today, the moon goddess, in her flowing robes and carrying the moon rabbit, sometimes flies on mooncake boxes.

The moon was a female goddess – Selene – in ancient Greece, and the Romans called her Luna. She was mentioned as early as the eighth century BC by the Greek poet Hesiod, who called her "bright Selene" in *Theogony*, and her name comes from the word for "light." Her Homeric Hymn tells us she is "long-winged." Her arms are white. Her hair shines. The flying horses that power her silver chariot let her drive the heavenly highway at full speed, and they appear with her on ancient ceramics. Often there is a crescent on her head. She is not just a symbol of rhythmic regeneration. She's the visible, divine, celestial incarnation of ever-changing growth and decline.

The Maya also made a goddess out of the moon, and in the Inca origin myth, the moon was the wife of the sun and the mother of the mythic ancestor couple. The Inca called the moon Quilla and used the same word for the month. The Maya frequently accompanied images of the moon goddess with a *U* symbol that likely references a crescent. Colonial images of the Inca moon goddess place her face in the sky on a full disk.

The goddess Coyolxauhqui has lunar attributes in Aztec myth. In a similar way, the Greek goddess Artemis, known to Rome as Diana, was not the real moon but a goddess affiliated with the moon. Artemis is represented as a huntress in ancient statues and on vases. Coyolxauhqui is portrayed as a decapitated head or a dismembered body.

The fertility cycle of women roughly matches the length of the moon's monthly transformation but is not universally synchronized with the moon's phases. The similarity of the two periods reinforced the symbolic connection between women and the moon in European folklore, and in our time, broad assertions about the femininity of the moon are common, particularly in astrology and in Pop-goddess literature. According to Barbara G. Walker's *The Woman's Dictionary of Symbols & Sacred Objects* (1988), the moon's "apparent connection with women's cycles of 'lunar blood,' which was supposed to give life to every human being in the womb," turned the moon into "the prime symbol of the Mother Goddess everywhere."

Responding to the Moon

There is no illusion in moonlight. It is a power over the darkness of night, and although the moon doesn't really drive the seasons, its behavior let us keep track of them.

In addition, maritime peoples correctly saw a relationship between the moon and the tides. The moon's gravity is the primary cause of the tides, and in responding to the tides, some living things respond to the power of the moon. At full moon some corals dramatically "bloom" with the release of a cloud of eggs and sperm.

The moon's genuine coercion of the oceans, the marine activity modulated by the moon, and the coincidental match of the human ovulation cycle with the period of the moon's phases all probably encourage magical thinking about the moon's physical effect on us. We are often told by police that more intoxication, accidents, violence, and crimes coincide with full moon, and firefighters say the full moon and false alarms are fast friends. Emergency rooms are supposed to be busier when the moon is full, and the full moon gets blamed for more disturbances in psychiatric wards – we still use the word "lunatic" to describe mental illness today. It's also claimed more babies are born at full moon. None of this is true. Selective memory is the force at work here, not the moon.

Most people understand the moon exerts a gravitational force responsible for the tides and figure it should pull on us, too, given that we are 50- to 60-percent water. It does, but the force of the moon on each of us is infinitesimal. The moon's direct force on us is about 32,000 times less than our weight, and that's roughly the same all the time regardless of the moon's phase or whether it happens to be in the sky. The moon's tidal force is even less, about 30 trillion times less than our weight. For an average person, the force is about 3 trillionths of a pound. The Louisiana Museum of Modern Art catalogue for this exhibit understandably has a greater physical impact on the reader than does the moon.

Changing the Moon

Conspicuous, useful, and powerful, the moon became personified and divine. For the Greeks and Romans, the moon was not only a goddess, it was a supernatural realm where gods and spirits live. Pythagoras of Samos, in the sixth and fifth centuries BC, regarded the moon as a haven for blessed souls after death and allegedly lodged them on the moon's far side. The Greek philosopher Aristotle regarded the moon as part of a cosmic system, and in ancient Rome, it also became an exotic destination.

Through antiquity and the centuries that followed, the moon remained a transcendental, celestial domain, but it was also tangible, physical, and solid. It was a gate to the Beyond, but literary dream voyages also carried heroes to its surface.

Lucian of Samosata, in the second century AD, imagined traveling there in his fictional *True History*. Although the moon wasn't exactly a world, it was a place. It had distinctive features – dark areas, or smudges – across its face, their nature unknown.

Ancient recognition of the marks on the moon is evident, however, in an Inuit tale that identifies the moon's darker areas as smudges of ash the moon applied to his face as a disguise that would fool his sister, the sun. In many different cultures around the globe, people see a rabbit in the moon's dark patches, and the Man in the Moon, however he is configured by people below, is something some eyes see in the mix of the moon's dark and light terrain. Some emblems of the moon's countenance – like the moon pictures in ancient Chinese tombs – reflect this perception of the lunar disk, but those pictures of the moon are not realistic portraits. There are rabbits and toads in the moon disks, not dark spots.

Any naturalistic depiction of the moon ought to include its mottling of dark and light, but naturalistic representation of the moon is not really common in antiquity. The moon usually is shown instead as an emblematic crescent or disk. A Seleucid era (301-164 BC) engraving from Mesopotamia portrays the moon with a man, perhaps the god Sin, getting the best of a lion in the disk and could be a stylized image of the moon's real face. The face that appears in a moon disk with Thessalian witches on a Greek vase from the second century BC may also be a stylized image of the actual moon, but these are mannered conventions, not realistic drawings. Even if the shapes in the lunar disks roughly conform to what's on the moon, neither of these accurately illustrates the moon.

The moon that signaled our seasons, inspired our calendars, turned into gods of cyclical renewal, and put werewolves on the prowl itself changed after tens of thousands of years of reliable phases and became something else in the seventeenth century when Galileo pointed a telescope at the sky and changed everything.

Romancing the Moon

We may have lost some of our romantic notions about the moon, but we've had no trouble conjuring up more moon magic. Despite modest domestication, the moon is still well outside ordinary human experience and is so hard to fathom that we tolerate in social discourse the absurd notion that we never went there at all. Disregarding the physical evidence of the chemically distinctive samples brought back from the moon and all of the authentic pictures, moon-landing-conspiracy believers insist the landings are all products of a Hollywood movie studio, concocted by a cynical and exploitive government operating through NASA.

The persistence of the moon-landing conspiracy in mass media does indicate the moon can still hold our hearts hostage with unexpected mystery. The moon was not diminished by the Space Program. It was, in fact, re-enchanted with new meanings, some silly and others profound. Moon-landing deniers and Apollo astronauts aren't the only ones who found one kind of magic or another in the moon. We all have. The moon is more popular than ever. It commands Internet headlines. It trends on Twitter. Blue moons, blood moons, and supermoons now return reliably to the public spotlight whenever the calendar, the moon's phases, and its orbital circumstances call them to order. All three of these lunar assignments converged in the 31 January 2018 total lunar eclipse, which was proclaimed the "Super Blue Blood Moon."

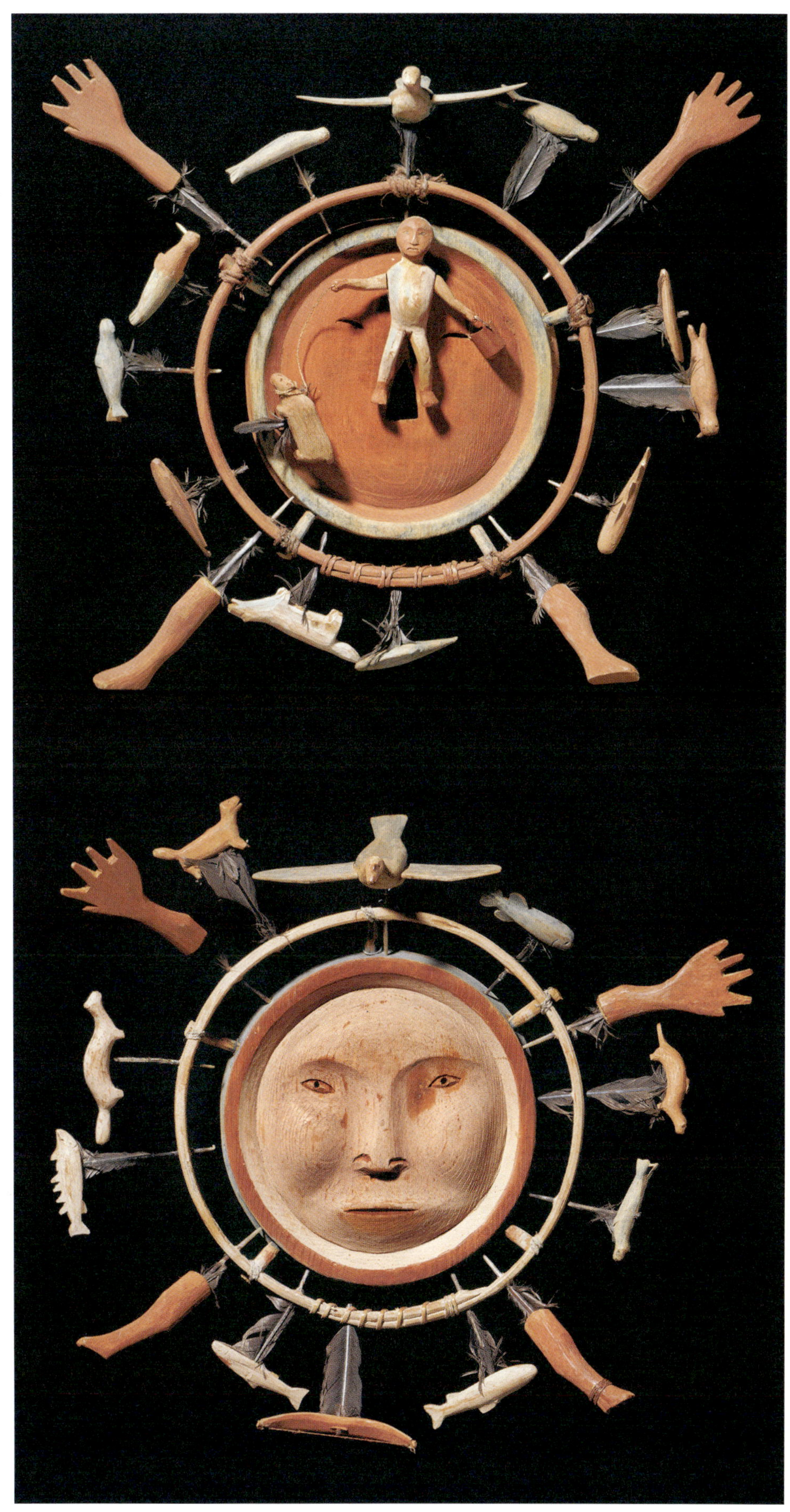

cat. 191
Moonlight mask from
Ninuvak Island

cat. 192
Sun mask

Of course, the moon turns dark in full eclipse, but it is not exactly super, blue, or bloody. The term *supermoon*, however, is now used to identify a full moon that occurs when the moon is near perigee, the moon's nearest monthly approach to the earth. The moon then commands interest because it is bigger – about six to seven percent – and brighter than the usual full moon, but the eye and brain are not able to discern the difference.

There is nothing astronomically official in the supermoon designation. It was invented by Richard Nolle, an astrologer, in 1979. Independently and arbitrarily, he decided the full moon is super whenever it is also within 90 percent of the distance of perigee. The blue moon is not the work of a superlative-seeking astrologer but is instead modern fabricated folklore.

The term *blue moon* is as familiar as a popular song, written in 1934 and recorded by everyone from Billie Holiday to Bob Dylan, and as colorful as a common proverb about the rarity of an event. The notion of an improbable blue moon can be tracked back to the sixteenth century, but in this era a blue moon has come to mean the second full moon in a calendar month. The modern blue moon was accidentally invented in 1946 in an article in *Sky & Telescope* magazine but remain ignored until the 1980s, when it was promoted on an astronomical radio show and seized by a popular game for home and family.

The "blood moon" is an even more recent affectation. Although the copper-red color of a moon in total eclipse was noted in antiquity, no one routinely called a total lunar eclipse a blood moon until 2008. Anticipating the 2014/2015 series of four total lunar eclipses, two American evangelical ministers, Reverend Mark Biltz and Reverend John Hagee, christened them all blood moons and predicted an imminent Apocalypse. The Apocalypse failed to arrive on schedule, but the damage was done. Now every total lunar eclipse is a blood moon, including the eclipse on 31 January 2018.

Off-color or not, the moon now shows up on smartphones all over the world. You don't need an almanac to know what the moon's doing. People still look at it, especially when media, social or otherwise, direct our attention to it. Many are persuaded the supermoon is really super because they have watched for the rising of that moon, and when they saw it, perched on the horizon, it looked huge. Every moon on the horizon looks huge, however. It's an optical illusion named for the moon – the "moon illusion," and for reasons only partially understood, the eye and the brain make the moon, already eye-catching among the mountains, trees, and other features in the near distance, seem to be much larger than it is but always as evocative as ever. Still conferring illusions, the moon still looms large in everyone's eyes. Although most people no longer follow it in detail and some are surprised to spot it in the daytime sky, it is a fixture in our hearts, for we keep finding ways to acknowledge it. In any phase, it commands attention, looks pretty in the sky, enriches the landscape, and adds light to the night. It seems to have meaning.

Further Reading

Jules Cashford, *The moon: myth and image*, Four Walls Eight Windows, 2003.
E.C. Krupp, "Beneath the moon, under the sun," in *Sun/moon: myths, tales, and visions*, The Nature Company, 1994.
E.C. Krupp, *Beyond the blue horizon: myths and legends of the sun, moon, stars, and planets*, HarperCollins Publishers, 1991.
E.C. Krupp, *Echoes of the ancient skies*, Harper & Row, Publishers, Inc., 1983.
E.C. Krupp, "Into the blue," in *Heavenly discourses* (ed. Nicholas Campion), Sophia Centre Press, 2016.
E.C. Krupp, "Moon maids," *Griffith Observer*, December, 1988.
E.C. Krupp, *The moon and you*, Macmillan Publishing Company, 1993.
Scott L. Montgomery, *The moon and the western imagination*, The University of Arizona Press, 1999.
Wernher von Braun, Silvio A. Bedini, and Fred L. Whipple, *Moon: man's greatest adventure*, Harry N. Abrams, Inc., Publishers, 1970.
Edgar Williams, *Moon: nature and culture*, Reaktion Books, 2014.

Jorge Luis Borges

The Moon

To María Kodama

There is such loneliness in that gold.
The moon of the nights is not the moon
Whom the first Adam saw. The long centuries
Of human vigil have filled her
With ancient lament. Look at her. She is your mirror.

Translated by Willis Barnstone. From *Borges: Selected Poems*, Penguin Books, 1999. This poem by Argentinian Jorge Luis Borges (1899-1986) is from 1976. Five years earlier Borges wrote '1971' about the moon landing, which begins "Two men walked on the surface of the moon./ Others will, later. What are words to do?/ And what of the dreams and fashionings of art / before this real, almost unreal, event?" Another poetic response to the 1969 Apollo landings is British W.H. Auden's 'Moon Landing': "Worth seeing? Mneh! I once rode through a desert/ and was not charmed"

Sappho

stars around the beautiful moon
hide back their luminous form
whenever all full she shines
        on the earth

        silvery

Moon has set
and Pleiades: middle
night, the hour goes by,
alone I die.

Translated by Anne Carson. From *If Not, Winter*, Virago, 2003. Two fragments, 34 and 168B, by the most famous female poet of antiquity, Sappho (c. 620-570). Recently scientists dated the second fragment to 570 BC on the basis of Sappho's description of the positions of the celestial bodies: i.e. that the moon appeared at the same time as the star cluster Pleiades (also known as the Seven Sisters) around midnight over the Greek island Lesbos, where Sappho lived

The Moon was but a Chin of Gold

The Moon was but a Chin of Gold
A Night or two ago –
And now she turns Her perfect Face
Upon the World below –

Her Forehead is of Amplest Blonde –
Her Cheek – a Beryl hewn –
Her Eye unto the Summer Dew
The likest I have known –

Her Lips of Amber never part –
But what must be the smile
Upon Her Friend she could confer
Were such Her Silver Will –

And what a privilege to be
But the remotest Star –
For Certainty She take Her Way
Beside Your Palace Door –

Her Bonnet is the Firmament –
The Universe – Her Shoe –
The Stars – the Trinkets at Her Belt –
Her Dimities – of Blue.

This poem from around 1863 is one of the several lunar poems written by American Emily Dickinson (1830-1886). In a letter to her sister in Baltimore, Dickinson also described the "sweet silver moon": "She looks like a fairy tonight, sailing around the sky in a little silver gondola with stars for gondoliers. I asked her to let me ride a little while ago – and told her I would *get out* when she got as far as Baltimore, but she only smiled to herself and went sailing on. I think she was quite ungenerous – but I have learned the lesson and shant ever ask her again"

Drinking Alone by Moonlight

A cup of wine, under the flowering trees;
I drink alone, for no friend is near.
Raising my cup I beckon the bright moon,
For he, with my shadow, will make three men.
The moon, alas, is no drinker of wine;
Listless, my shadow creeps about at my side.
Yet with the moon as friend and the shadow as slave
I must make merry before the Spring is spent.
To the songs I sing the moon flickers her beams;
In the dance I weave my shadow tangles and breaks.
While we were sober, three shared the fun;
Now we are drunk, each goes his way.
May we long share our odd, inanimate feast,
And meet at last on the Cloudy River of the sky.

Translated by Arthur Waley. This poem, from around 743, by Chinese Li Po (701-762) is one of the most famous poems from the Tang Dynasty. The image of the poet drinking alone in moonlight is recurrent all through Li Po's work. The American modernist Ezra Pound's 'Epitaphs' says about Li Po: "He tried to embrace a moon/ In the Yellow River", referring to the legend that the poet drowned while drunk, trying to embrace the reflection of the moon in the water

To The Moon

Art thou pale for weariness
Of climbing heaven and gazing on the earth,
Wandering companionless
Among the stars that have a different birth,
And ever changing, like a joyless eye
That finds no object worth its constancy?

The Waning Moon

And like a dying lady, lean and pale,
Who totters forth, wrapped in a gauzy veil,
Out of her chamber, led by the insane
And feeble wanderings of her fading brain,
The moon arose up in the murky East,
A white and shapeless mass –

These two poems from 1820 by the English Romantic poet Percy Bysshe Shelley (1792-1822) continue a long tradition in western poetry of attributing female qualities to the moon. Shakespeare too used the image of the moon as an old woman in *A Midsummer Night's Dream*: "how slow/ This old moon wanes! she lingers my desires,/ Like to a step-dame or a dowager." And in *Astrophel and Stella* from 1582 the Elizabethan poet Sir Philip Sidney famously addressed the paleness of the moon: "With how sad steps, o Moone, thou climbs't the skies,/ How silently, and with how wanne a face"

The Moon and the Yew Tree

This is the light of the mind, cold and planetary.
The trees of the mind are black. The light is blue.
The grasses unload their griefs at my feet as if I were God,
Prickling my ankles and murmuring of their humility.
Fumy, spiritous mists inhabit this place
Separated from my house by a row of headstones.
I simply cannot see where there is to get to.

The moon is no door. It is a face in its own right,
White as a knuckle and terribly upset.
It drags the sea after it like a dark crime; it is quiet
With the O-gape of complete despair. I live here.
Twice on Sunday, the bells startle the sky –
Eight great tongues affirming the Resurrection.
At the end, they soberly bong out their names.

The yew tree points up. It has a Gothic shape.
The eyes lift after it and find the moon.
The moon is my mother. She is not sweet like Mary.
Her blue garments unloose small bats and owls.
How I would like to believe in tenderness –
The face of the effigy, gentled by candles,
Bending, on me in particular, its mild eyes.

I have fallen a long way. Clouds are flowering
Blue and mystical over the face of the stars.
Inside the church, the saints will be all blue,
Floating on their delicate feet over the cold pews,
Their hands and faces stiff with holiness.
The moon sees nothing of this. She is bald and wild.
And the message of the yew tree is blackness – blackness and silence.

'The Moon and the Yew Tree' is one of several lunar poems from the poetry collection *Ariel* by American Sylvia Plath (1933-1963), written shortly before she committed suicide, published posthumously

Walt Whitman

Look Down Fair Moon

Look down fair moon and bathe this scene;
Pour softly down night's nimbus floods on faces ghastly, swollen, purple,
On the dead on their back with their arms tos'd wide,
Pour down your unstinted nimbus sacred moon.

In this poem from the classic *Leaves of Grass* (1855) by American Walt Whitman (1819-1892) the light of the moon shines on dead soldiers. The moon is an ancient symbol of death, because it 'dies' in the sky in each cycle. In the famous poem ballad by the English Romantic poet William Wordsworth 'Strange Fits of Passion Have I Known' from 1815, the moon's disappearance from the sky leads to the recognition that the beloved Lucy must also die one day

Moonrise

And who has seen the moon, who has not seen
Her rise from out the chamber of the deep
Flushed and grand and naked, as from the chamber
Of finished bridegroom, seen her rise and throw
Confession of delight upon the wave,
Littering the waves with her own superscription
Of bliss, till all her lambent beauty shakes towards us
Spread out and known at last: and we are sure
That beauty is a thing beyond the grave,
That perfect, bright experience never falls
To nothingness, and time will dim the moon
Sooner than our full consummation here
In this odd life will tarnish or pass away.

In this poem from 1912 by the English writer D.H. Lawrence (1885-1930) the moon is described with erotic undertones. In another of Lawrence's moon-themed poems, 'Red Moon-Rise', the ascending moon is compared to a birth: "And then, from between shut lips of darkness, red/ As if from the womb the slow moon rises, as if the twin-walled darkness had bled/ In a new night-spasm of birth"

The moonlight seen through the tall branches
Is more, say all the poets,
Than the moonlight seen through the tall branches.

But for me, oblivious to what I think,
The moonlight seen through the tall branches,
Besides its being
The moonlight seen through the tall branches,
Is its not being more
Than the moonlight seen through the tall branches.

Translated by Richard Zenith. From *A Little Larger Than the Entire Universe. Selected Poems*, Penguin, 2006. This poem from 1914 by the Portuguese Fernando Pessoa (1888-1935) is a clear showdown with the long tradition of poets who have used the moon as a metaphor. Another modern approach to moonlight poetry is Yoko Ono's instruction piece "Steal a moon on the water with a bucket. Keep stealing until no moon is seen on the water" from *Grapefruit* (1965)

Italo Calvino

The Distance of the Moon

At one time, according to Sir George H. Darwin, the Moon was very close to the Earth. Then the tides gradually pushed her far away: the tides that the Moon herself causes in the Earth's waters, where the Earth slowly loses energy.

Translated by William Weaver. From *The Distance of the Moon*, Penguin, 2018. The short story from *Cosmicomics* (1967) by Italian Italo Calvino (1923-85) is played out at an early stage of the history of the solar system, when the moon was closer to the earth. The magical story takes as its starting point the scientific discovery that each year the moon distances itself from earth

How well I know! – *old Qfwfq cried* – the rest of you can't remember, but I can. We had her on top of us all the time, that enormous Moon: when she was full – nights as bright as day, but with a butter-coloured light – it looked as if she were going to crush us; when she was new, she rolled around the sky like a black umbrella blown by the wind; and when she was waxing, she came forward with her horns so low she seemed about to stick into the peak of a promontory and get caught there. But the whole business of the Moon's phases worked in a different way then: because the distances from the Sun were different, and the orbits, and the angle of something or other, I forget what; as for eclipses, with Earth and Moon stuck together the way they were, why, we had eclipses every minute: naturally, those two big monsters managed to put each other in the shade constantly, first one, then the other.

Orbit? Oh, elliptical, of course: for a while it would huddle against us and then it would take flight for a while. The tides, when the Moon swung closer, rose so high nobody could hold them back. There were nights when the Moon was full and very, very low, and the tide was so high that the Moon missed a ducking in the sea by a hair's breadth; well, let's say a few yards anyway. Climb up on the Moon? Of course we did. All you had to do was row out to it in a boat and, when you were underneath, prop a ladder against her and scramble up.

The spot where the Moon was lowest, as she went by, was off the Zinc Cliffs. We used to go out with those little rowing boats they had in those days, round and flat, made of cork. They held quite a few of us: me, Captain Vhd Vhd, his wife, my deaf cousin, and sometimes little Xlthlx – she was twelve or so at that time. On those nights the water was very calm, so silvery it looked like mercury, and the fish in it, violet-coloured, unable to resist the Moon's attraction, rose to the surface, all of them, and so did the octopuses and the saffron medusas. There was always a flight of tiny creatures – little crabs, squid, and even some weeds, light and filmy, and coral plants – that broke from the sea and ended up on the Moon, hanging down from that lime-white ceiling, or else they stayed in mid-air, a phosphorescent swarm we had to drive off, waving banana leaves at them.

This is how we did the job: in the boat we had a ladder: one of us held it, another climbed to the top, and a third, at the oars, rowed until we were right under the Moon; that's why there had to be so many of us (I only mentioned the main ones). The man at the top of the ladder, as the boat approached the Moon, would become scared and start shouting: 'Stop! Stop! I'm going to bang my head!' That was the impression you had, seeing her on top of you, immense, and all rough with sharp spikes and jagged, saw-tooth edges. It may be different now, but then the Moon, or rather the bottom, the underbelly of the Moon, the part that passed closest to the Earth and almost scraped it, was covered with a crust of sharp scales. It had come to resemble the belly of a fish, and the smell too, as I recall, if not downright fishy, was faintly similar, like smoked salmon.

In reality, from the top of the ladder, standing erect on the last rung, you could just touch the Moon if you held your arms up. We had taken the measurements carefully (we didn't yet suspect that she was moving away from us); the only thing you had to be very careful about was where you put your hands. I always chose a scale that seemed fast (we climbed up in groups of five or six at a time), then I would cling first with one hand, then with both, and immediately I would feel ladder and boat drifting away from below me, and the motion of the Moon would tear me from the Earth's attraction. Yes, the Moon was so strong that she pulled you up; you realized this the moment you passed from one to the other: you had to swing up abruptly, with a kind of somersault, grabbing the scales, throwing your legs over your head, until your feet were on the Moon's surface. Seen from the Earth, you looked as if you were hanging there with your head down, but for you, it was the normal position, and the only odd thing was that when you raised your eyes you saw the sea above you, glistening, with the boat and the others upside down, hanging like a bunch of grapes from the vine.

My cousin, the Deaf One, showed a special talent for making those leaps. His clumsy hands, as soon as they touched the lunar surface (he was always the first to jump up from the ladder), suddenly became deft and sensitive. They found immediately the spot where he could hoist himself up; in fact just the pressure of his palms seemed enough to make him stick to the satellite's crust. Once I even thought I saw the Moon come towards him, as he held out his hands.

He was just as dextrous in coming back down to Earth, an operation still more difficult. For us, it consisted in jumping, as high as we could, our arms upraised (seen from the Moon, that is, because seen from the Earth it looked more like a dive, or like swimming downwards, arms at our sides), like jumping up from the Earth in other words, only now we were without the ladder, because there was nothing to prop it against on the Moon. But instead of jumping with his arms out, my cousin bent towards the Moon's surface, his head down as if for a somersault, then made a leap, pushing with his hands. From the boat we watched him, erect in the air as if he were supporting the Moon's enormous ball and were tossing it, striking it with his palms; then, when his legs came within reach, we managed to grab his ankles and pull him down on board.

Now, you will ask me what in the world we went up on the Moon for; I'll explain it to you. We went to collect the milk, with a big spoon and a bucket. Moon-milk was very thick, like a kind of cream cheese. It formed in the crevices between one scale and the next, through the fermentation of various bodies and substances of terrestrial origin which had flown up from the prairies and forests and lakes, as the Moon sailed over them. It was composed chiefly of vegetal juices, tadpoles, bitumen, lentils, honey, starch crystals, sturgeon eggs, moulds, pollens, gelatinous matter, worms, resins, pepper, mineral salts, combustion residue. You had only to dip the spoon under the scales that covered the Moon's scabby terrain, and you brought it out filled with that precious muck. Not in the pure state, obviously; there was a lot of refuse. In the fermentation (which took place as the Moon passed over the expanses of hot air above the deserts) not all the bodies melted; some remained stuck in it: fingernails and cartilage, bolts, sea horses, nuts and peduncles, shards of crockery, fish-hooks, at times even a comb. So this paste, after it was collected, had to be refined, filtered. But that wasn't the difficulty: the hard part was transporting it down to the Earth. This is how we did it: we hurled each spoonful into the air with both hands, using the spoon as a catapult. The cheese flew, and if we had thrown it hard enough, it stuck to the ceiling, I mean the surface of the sea. Once there, it floated, and it was easy enough to pull it into the boat. In this operation, too, my deaf cousin displayed a special gift; he had strength and a good aim; with a single, sharp throw, he could send the cheese straight into a bucket we held up to him from the boat. As for me, I occasionally misfired; the contents of the spoon would fail to overcome the Moon's attraction and they would fall back into my eye.

I still haven't told you everything about the things my cousin was good at. That job of extracting lunar milk from the Moon's scales was child's play to him: instead of the spoon, at times he had only to thrust his bare hand under the scales, or even one finger. He didn't proceed in any orderly way, but went to isolated places, jumping from one to the other, as if he were playing tricks on the Moon, surprising her, or perhaps tickling her. And wherever he put his hand, the milk spurted out as if from a nanny goat's teats. So the rest of us had only to follow him and collect with our spoons the substance that he was pressing out, first here, then there, but always as if by chance, since the Deaf One's movements seemed to have no clear, practical sense. There were places, for example, that he touched merely for the fun of touching them: gaps between two scales, naked and tender folds of lunar flesh. At times my cousin pressed not only his fingers but – in a carefully gauged leap – his big toe (he climbed on to the Moon barefoot) and this seemed to be the height of amusement for him, if we could judge by the chirping sounds that came from his throat as he went on leaping.

The soil of the Moon was not uniformly scaly, but revealed irregular bare patches of pale, slippery clay. These soft areas inspired the Deaf One to turn somersaults or to fly almost like a bird, as if he wanted to impress his whole body into the Moon's pulp. As he ventured further in this way, we lost sight of him at one point. On the Moon there were vast areas we had never had any reason or curiosity to explore, and that was where my cousin vanished; I had suspected that all those somersaults and nudges he indulged in before our eyes were only a preparation, a prelude to something secret meant to take place in the hidden zones.

We fell into a special mood on those nights off the Zinc Cliffs: gay, but with a touch of suspense, as if inside our skulls, instead of the brain, we felt a fish, floating, attracted by the Moon. And so we navigated, playing and singing. The Captain's wife played the harp; she had very long arms, silvery as eels on those nights, and armpits as dark and mysterious as sea urchins; and the sound of the harp was sweet and piercing, so sweet and piercing it was almost unbearable, and we were forced to let out long cries, not so much to accompany the music as to protect our hearing from it.

Transparent medusas rose to the sea's surface, throbbed there a moment, then flew off, swaying towards the Moon. Little Xlthlx amused herself by catching them in midair, though it wasn't easy. Once, as she stretched her little arms out to catch one, she jumped up slightly and was also set free. Thin as she was, she was an ounce or two short of the weight necessary for the Earth's gravity to overcome the Moon's attraction and bring her back: so she flew up among the medusas, suspended over the sea. She took fright, cried, then laughed and started playing, catching shellfish and minnows as they flew, sticking some into her mouth and chewing them. We rowed hard, to keep up with the child: the Moon ran off in her ellipse, dragging that swarm of marine fauna through the sky, and a train of long, entwined seaweeds, and Xlthlx hanging there in the midst. Her two wispy braids seemed to be flying on their own, outstretched towards the Moon; but all the while she kept wriggling and kicking at the air, as if she wanted to fight that influence, and her socks – she had lost her shoes in the fight – slipped off her feet and swayed, attracted by the Earth's force. On the ladder, we tried to grab them.

The idea of eating the little animals in the air had been a good one; the more weight Xlthlx gained, the more she sank towards the Earth; in fact, since among those hovering bodies hers was the largest, molluscs and seaweeds and plankton began to gravitate about her, and soon the child was covered with siliceous little shells, chitinous carapaces and fibres of sea plants. And the further she vanished into that tangle, the more she was freed of the Moon's influence, until she grazed the surface of the water and sank into the sea.

We rowed quickly, to pull her out and save her: her body had remained magnetized, and we had to work hard to scrape off all the things encrusted on her. Tender corals were wound about her head, and every time we ran the comb through her hair there was a shower of crayfish and sardines; her eyes were sealed shut by limpets clinging to the lids with their suckers; squids' tentacles were coiled around her arms and her neck; and her little dress now seemed woven only of weeds and sponges. We got the worst of it off her, but for weeks afterwards she went on pulling out fins and shells, and her skin, dotted with little diatoms, remained affected for ever, looking – to someone who didn't observe her carefully – as if it were faintly dusted with freckles.

This should give you an idea of how the influences of Earth and Moon, practically equal, fought over the space between them. I'll tell you something else: a body that descended to the Earth from the satellite was still charged for a while with lunar force and rejected the attraction of our world. Even I, big and heavy as I was: every time I had been up there, I took a while to get used to the Earth's up and its down, and the others would have to grab my arms and hold me, clinging in a bunch in the swaying boat while I still had my head hanging and my legs stretching up towards the sky.

'Hold on! Hold on to us!' they shouted at me, and in all that groping, sometimes I ended up by seizing one of Mrs Vhd Vhd's breasts, which were round and firm and the contact was good and secure and had an attraction as strong as the Moon's or even stronger, especially if I managed, as I plunged down, to put my other arm around her hips, and with this I passed back into our world and fell with a thud into the bottom of the boat, where Captain Vhd Vhd brought me around, throwing a bucket of water in my face.

This is how the story of my love for the Captain's wife began, and my suffering. Because it didn't take me long to realize whom the lady kept looking at insistently: when my cousin's hands clasped the satellite, I watched Mrs Vhd Vhd, and in her eyes I could read the thoughts that the deaf man's familiarity with the Moon were arousing in her; and when he disappeared in his mysterious lunar explorations, I saw her become restless, as if on pins and needles, and then it was all clear to me, how Mrs Vhd Vhd was becoming jealous of the Moon and I was jealous of my cousin. Her eyes were made of diamonds, Mrs Vhd Vhd's; they flared when she looked at the Moon, almost challengingly, as if she were saying: 'You shan't have him!' And I felt like an outsider.

The one who least understood all of this was my deaf cousin. When we helped him down, pulling him – as I explained to you – by his legs, Mrs Vhd Vhd lost all her self-control, doing everything she could to take his weight against her own body, folding her long silvery arms around him; I felt a pang in my heart (the times I clung to her, her body was soft and kind, but not thrust forward, the way it was with my cousin), while he was indifferent, still lost in his lunar bliss.

I looked at the Captain, wondering if he also noticed his wife's behaviour; but there was never a trace of any expression on that face of his, eaten by brine, marked with tarry wrinkles. Since the Deaf One was always the last to break away from the Moon, his return was the signal for the boats to move off. Then, with an unusually polite gesture, Vhd Vhd picked up the harp from the bottom of the boat and handed it to his wife. She was obliged to take it and play a few notes. Nothing could separate her more from the Deaf One than the sound of the harp. I took to singing in a low voice that sad song that goes: 'Every shiny fish is floating, floating; and every dark fish is at the bottom, at the bottom of the sea ...' and all the others, except my cousin, echoed my words.

Every month, once the satellite had moved on, the Deaf One returned to his solitary detachment from the things of the world; only the approach of the full moon aroused him again. That time I had arranged things so it wasn't my turn to go up, I could stay in the boat with the Captain's wife. But then, as soon as my cousin had climbed the ladder, Mrs Vhd Vhd said: 'This time I want to go up there, too!'

This had never happened before; the Captain's wife had never gone up on the Moon. But Vhd Vhd made no objection, in fact he almost pushed her up the ladder bodily, exclaiming: 'Go ahead then!' and we all started helping her, and I held her from behind, felt her round and soft on my arms, and to hold her up I began to press my face and the palms of my hands against her, and when I felt her rising into the Moon's sphere I was heartsick at that lost contact, so I started to rush after her, saying: 'I'm going to go up for a while, too, to help out!'

I was held back as if in a vice. 'You stay here; you have work to do later,' the Captain commanded, without raising his voice.

At that moment each one's intentions were already clear. And yet I couldn't figure things out; even now I'm not sure I've interpreted it all correctly. Certainly the Captain's wife had for a long time been cherishing the desire to go off privately with my cousin up there (or at least to prevent him from going off alone with the Moon), but probably she had a still more ambitious plan, one that would have to be carried out in agreement with the Deaf One: she wanted the two of them to hide up there together and stay on the Moon for a month. But perhaps my cousin, deaf as he was, hadn't understood anything of what she had tried to explain to him, or perhaps he hadn't even realized that he was the object of the lady's desires. And the Captain? He wanted nothing better than to be rid of his wife; in fact, as soon as she was confined up there, we saw him give free rein to his inclinations and plunge into vice, and then we understood why he had done nothing to hold her back. But had he known from the beginning that the Moon's orbit was widening?

None of us could have suspected it. The Deaf One perhaps, but only he: in the shadowy way he knew things, he may have had a presentiment that he would be forced to bid the Moon farewell that night. This is why he hid in his secret places and reappeared only when it was time to come back down on board. It was no use for the Captain's wife to try to follow him: we saw her cross the scaly zone various times, length and breadth, then suddenly she stopped, looking at us in the boat, as if about to ask us whether we had seen him.

Surely there was something strange about that night. The sea's surface, instead of being taut as it was during the full moon, or even arched a bit towards the sky, now seemed limp, sagging, as if the lunar magnet no longer exercised its full power. And the light, too, wasn't the same as the light of other full moons; the night's shadows seemed somehow to have thickened. Our friends up there must have realized what was happening; in fact, they looked up at us with frightened eyes. And from their mouths and ours, at the same moment, came a cry: 'The Moon's going away!'

The cry hadn't died out when my cousin appeared on the Moon, running. He didn't seem frightened, or even amazed: he placed his hands on the terrain, flinging himself into his usual somersault, but this time after he had hurled himself into the air he remained suspended, as little Xlthlx had. He hovered a moment between Moon and Earth, upside down, then laboriously moving his arms, like someone swimming against a current, he headed with unusual slowness towards our planet.

From the Moon the other sailors hastened to follow his example. Nobody gave a thought to getting the Moon-milk that had been collected into the boats, nor did the Captain scold them for this. They had already waited too long, the distance was difficult to cross by now; when they tried to imitate my cousin's leap or his swimming, they remained there groping, suspended in mid-air. 'Cling together! Idiots! Cling together!' the Captain yelled. At this command, the sailors tried to form a group, a mass, to push all together until they reached the zone of the Earth's attraction: all of a sudden a cascade of bodies plunged into the sea with a loud splash.

The boats were now rowing to pick them up. 'Wait! The Captain's wife is missing!' I shouted. The Captain's wife had also tried to jump, but she was still floating only a few yards from the Moon, slowly moving her long, silvery arms in the air. I climbed up the ladder, and in a vain attempt to give her something to grasp I held the harp out towards her. 'I can't reach her! We have to go after her!' and I started to jump up, brandishing the harp. Above me the enormous lunar disc no longer seemed the same as before: it had become much smaller, it kept contracting, as if my gaze were driving it away, and the emptied sky gaped like an abyss where, at the bottom, the stars had begun multiplying, and the night poured a river of emptiness over me, drowned me in dizziness and alarm.

'I'm afraid,' I thought. 'I'm too afraid to jump. I'm a coward!' and at that moment I jumped. I swam furiously through the sky, and held the harp out to her, and instead of coming towards me she rolled over and over, showing me first her impassive face and then her backside.

'Hold tight to me!' I shouted, and I was already overtaking her, entwining my limbs with hers. 'If we cling together we can go down!' and I was concentrating all my strength on uniting myself more closely with her, and I concentrated my sensations as I enjoyed the fullness of that embrace. I was so absorbed I didn't realize at first that I was, indeed, tearing her from her weightless condition, but was making her fall back on the Moon. Didn't I realize it? Or had that been my intention from the very beginning? Before I could think properly, a cry was already bursting from my throat. 'I'll be the one to stay with you for a month!' Or rather, 'On you!' I shouted, in my excitement: 'On you for a month!' and at that moment our embrace was broken by our fall to the Moon's surface, where we rolled away from each other among those cold scales.

I raised my eyes as I did every time I touched the Moon's crust, sure that I would see above me the native sea like an endless ceiling, and I saw it, yes, I saw it this time, too, but much higher, and much more narrow, bound by its borders of coasts and cliffs and promontories, and how small the boats seemed, and how unfamiliar my friends' faces and how weak their cries! A sound reached me from nearby: Mrs Vhd Vhd had discovered her harp and was caressing it, sketching out a chord as sad as weeping.

A long month began. The Moon turned slowly around the Earth. On the suspended globe we no longer saw our familiar shore, but the passage of oceans as deep as abysses and deserts of glowing lapilli, and continents of ice, and forests writhing with reptiles, and the rocky walls of mountain chains gashed by swift rivers, and swampy cities, and stone graveyards, and empires of clay and mud. The distance spread a uniform colour over everything: the alien perspectives made every image alien; herds of elephants and swarms of locusts ran over the plains, so evenly vast and dense and thickly grown that there was no difference among them.

I should have been happy: as I had dreamed, I was alone with her, that intimacy with the Moon I had so often envied my cousin and with Mrs Vhd Vhd was now my exclusive prerogative, a month of days and lunar nights stretched uninterrupted before us, the crust of the satellite nourished us with its milk, whose tart flavour was familiar to us, we raised our eyes up, up to the world where we had been born, finally traversed in all its various expanse, explored landscapes no Earth-being had ever seen, or else we contemplated the stars beyond the Moon, big as pieces of fruit, made of light, ripened on the curved branches of the sky, and everything exceeded my most luminous hopes, and yet, and yet, it was, instead, exile.

I thought only of the Earth. It was the Earth that caused each of us to be that someone he was rather than someone else; up there, wrested from the Earth, it was as if I were no longer that I, nor she that She, for me. I was eager to return to the Earth, and I trembled at the fear of having lost it. The fulfilment of my dream of love had lasted only that instant when we had been united, spinning between Earth and Moon; torn from its earthly soil, my love now knew only the heart-rending nostalgia for what it lacked: a where, a surrounding, a before, an after.

This is what I was feeling. But she? As I asked myself, I was torn by my fears. Because if she also thought only of the Earth, this could be a good sign, a sign that she had finally come to understand me, but it could also mean that everything had been useless, that her longings were directed still and only towards my deaf cousin. Instead, she felt nothing. She never raised her eyes to the old planet, she went off, pale, among those wastelands, mumbling dirges and stroking her harp, as if completely identified with her temporary (as I thought) lunar state. Did this mean I had won out over my rival? No; I had lost: a hopeless defeat. Because she had finally realized that my cousin loved only the Moon, and the only thing she wanted now was to become the Moon, to be assimilated into the object of that extrahuman love.

When the Moon had completed its circling of the planet, there we were again over the Zinc Cliffs. I recognized them with dismay: not even in my darkest previsions had I thought the distance would have made them so tiny. In that mud puddle of the sea, my friends had set forth again, without the now-useless ladders; but from the boats rose a kind of forest of long poles; everybody was brandishing one, with a harpoon or a grappling hook at the end, perhaps in the hope of scraping off a last bit of Moon-milk or of lending some kind of help to us wretches up there. But it was soon clear that no pole was long enough to reach the Moon; and they dropped back, ridiculously short, humbled, floating on the sea; and in that confusion some of the boats were thrown off balance and overturned. But just then, from another vessel a longer pole, which till then they had dragged along on the water's surface, began to rise: it must have been made of bamboo, of many, many bamboo poles stuck one into the other, and to raise it they had to go slowly because – thin as it was – if they let it sway too much it might break. Therefore, they had to use it with great strength and skill, so that the wholly vertical weight wouldn't rock the boat.

Suddenly it was clear that the tip of that pole would touch the Moon, and we saw it graze, then press against the scaly terrain, rest there a moment, give a kind of little push, or rather a strong push that made it bounce off again, then come back and strike that same spot as if on the rebound, then move away once more. And I recognized, we both – the Captain's wife and I – recognized my cousin: it couldn't have been anyone else, he was playing his last game with the Moon, one of his tricks, with the Moon on the tip of his pole as if he were juggling with her. And we realized that his virtuosity had no purpose, aimed at no practical result, indeed you would have said he was driving the Moon away, that he was helping her departure, that he wanted to show her to her more distant orbit. And this, too, was just like him: he was unable to conceive desires that went against the Moon's nature, the Moon's course and destiny, and if the Moon now tended to go away from him, then he would take delight in this separation just as, till now, he had delighted in the Moon's nearness.

What could Mrs Vhd Vhd do, in the face of this? It was only at this moment that she proved her passion for the deaf man hadn't been a frivolous whim but an irrevocable vow. If what my cousin now loved was the distant Moon, then she too would remain distant, on the Moon. I sensed this, seeing that she didn't take a step towards the bamboo pole, but simply turned her harp towards the Earth, high in the sky, and plucked the strings. I say I saw her, but to tell the truth I only caught a glimpse of her out of the corner of my eye, because the minute the pole had touched the lunar crust, I had sprung and grasped it, and now, fast as a snake, I was climbing up the bamboo knots, pushing myself along with jerks of my arms and knees, light in the rarefied space, driven by a natural power that ordered me to return to the Earth, oblivious of the motive that had brought me here, or perhaps more aware of it than ever and of its unfortunate outcome; and already my climb up the swaying pole had reached the point where I no longer had to make any effort but could just allow myself to slide, head first, attracted by the Earth, until in my haste the pole broke into a thousand pieces and I fell into the sea, among the boats.

My return was sweet, my home refound, but my thoughts were filled only with grief at having lost her, and my eyes gazed at the Moon, for ever beyond my reach, as I sought her. And I saw her. She was there where I had left her, lying on a beach directly over our heads, and she said nothing. She was the colour of the Moon; she held the harp at her side and moved one hand now and then in slow arpeggios. I could distinguish the shape of her bosom, her arms, her thighs, just as I remember them now, just as now, when the Moon has become that flat, remote circle, I still look for her as soon as the first silver appears in the sky, and the more it waxes, the more clearly I imagine I can see her, her or something of her, but only her, in a hundred, a thousand different vistas, she who makes the Moon the Moon and, whenever she is full, sets the dogs to howling all night long, and me with them.

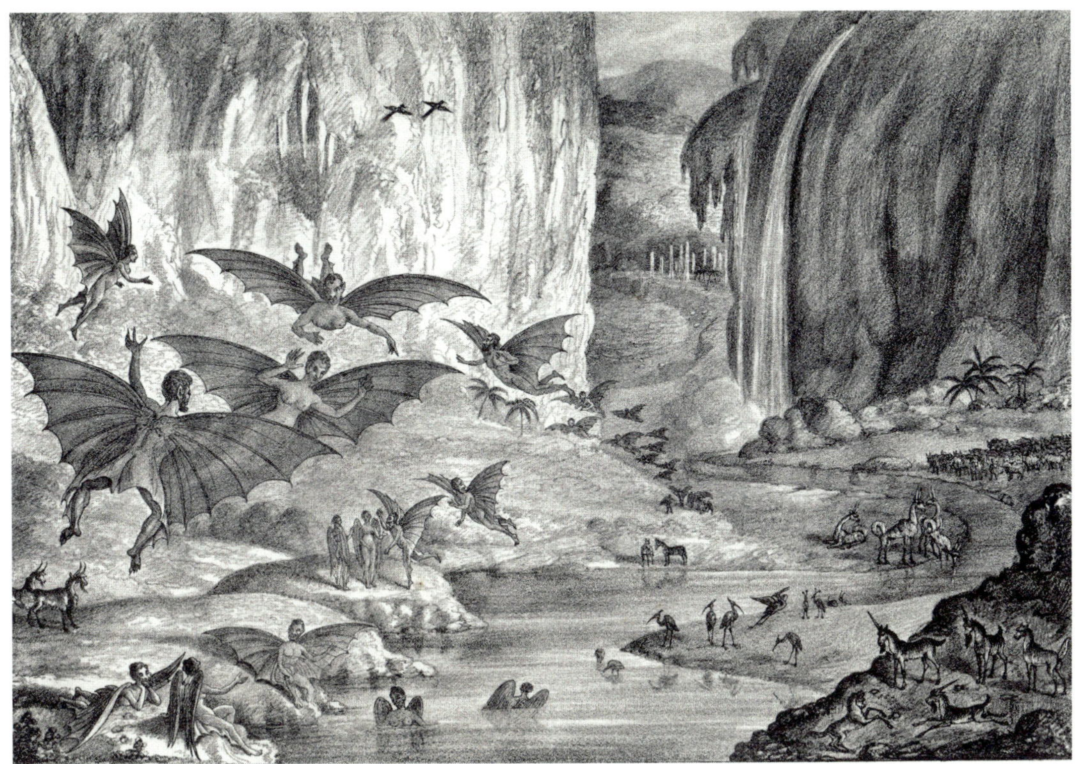

The British astronomer Sir John Herschel's observations of the moon inhabited by bat-like winged humans and unicorns in luxuriant landscapes. The story was later called the 'Great Moon Hoax'. Illustration from Edinburgh Journal of Science, 1835

cat. 61
Leopold Galluzzo's illustrations for the series of articles about the astronomer Sir John Herschel's discoveries of life on the moon – the 'Great Moon Hoax' – first published in The New York Sun in 1835

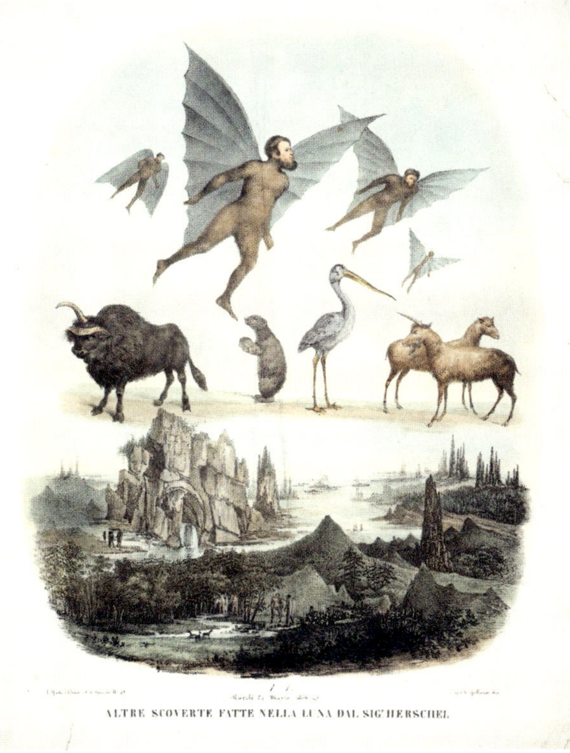

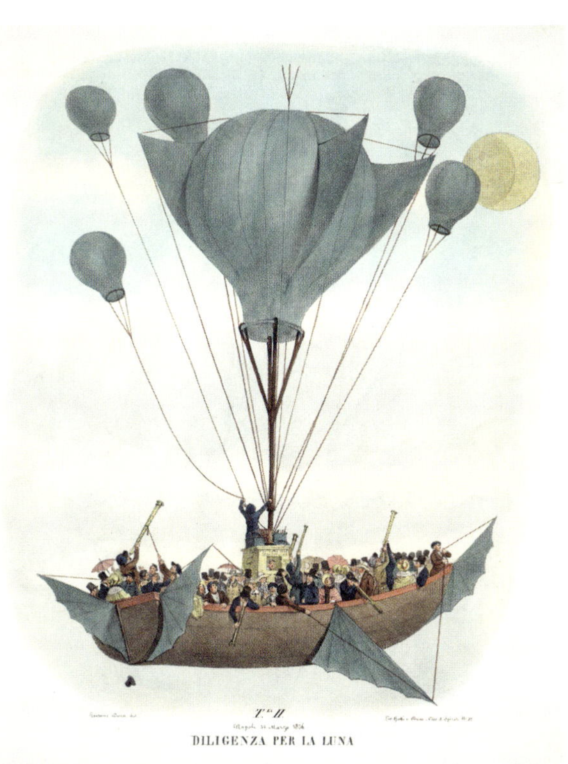

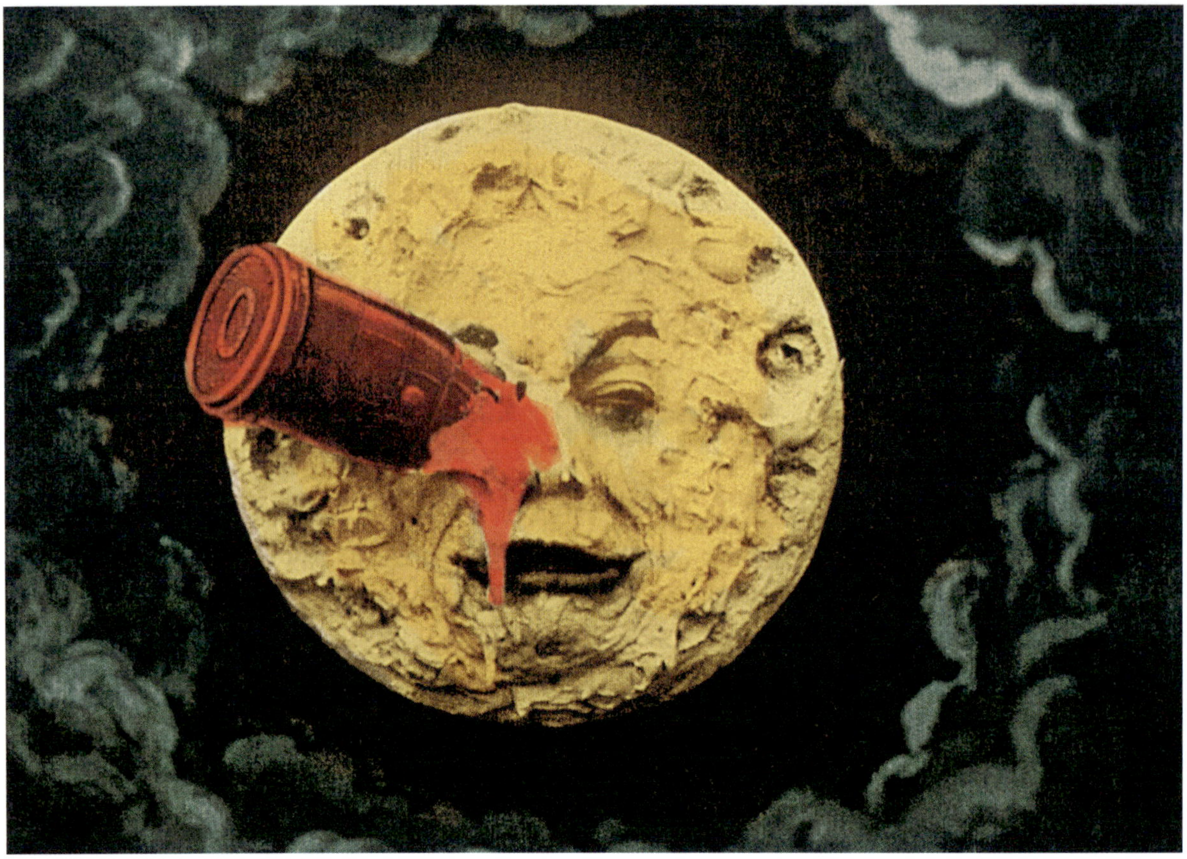

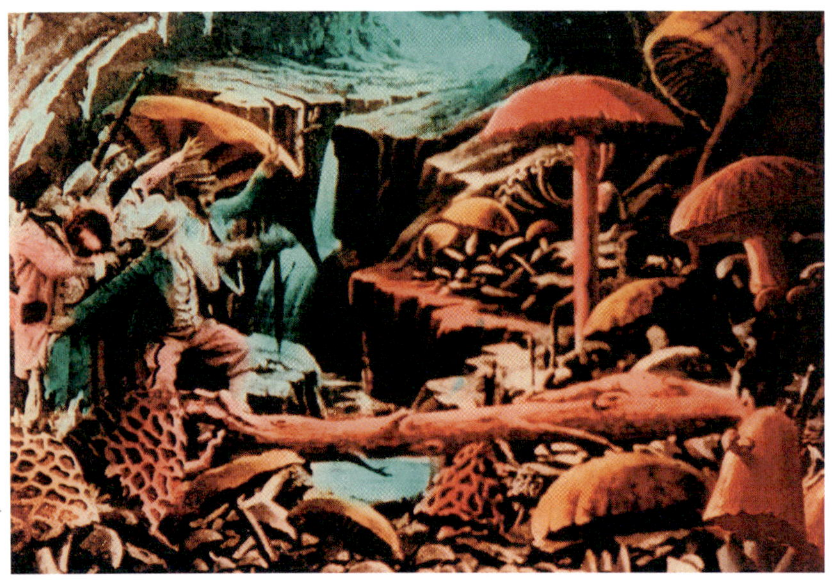

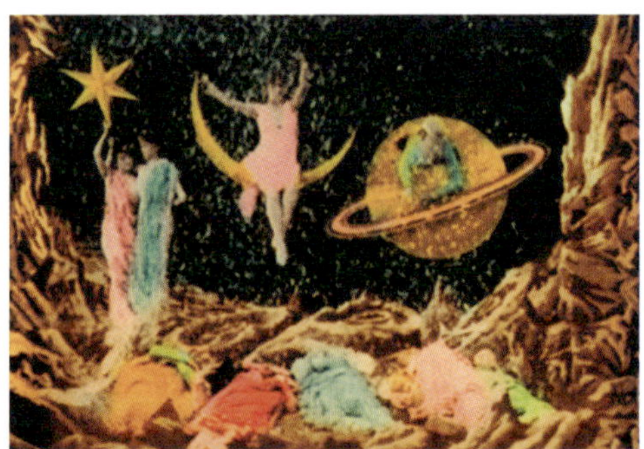

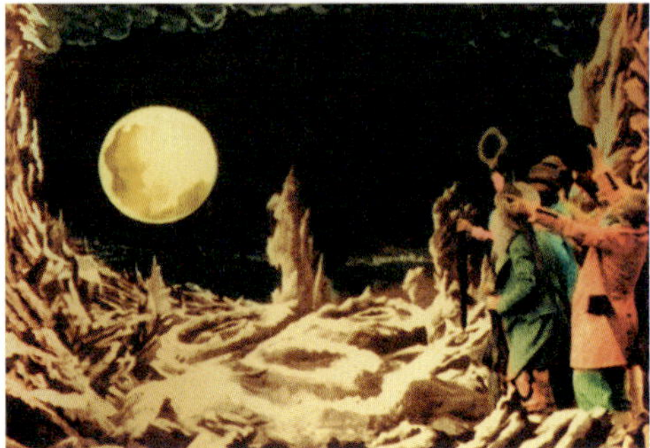

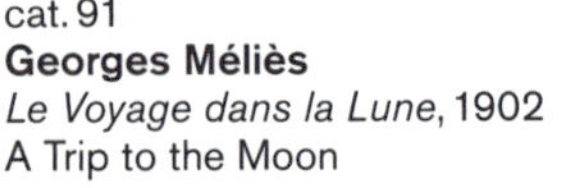

cat. 91
Georges Méliès
Le Voyage dans la Lune, 1902
A Trip to the Moon

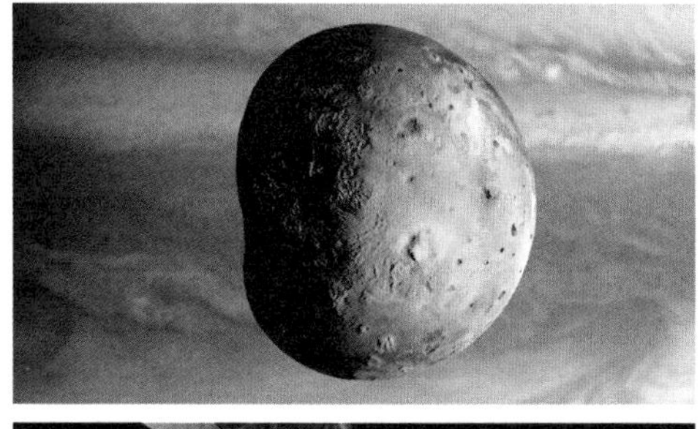

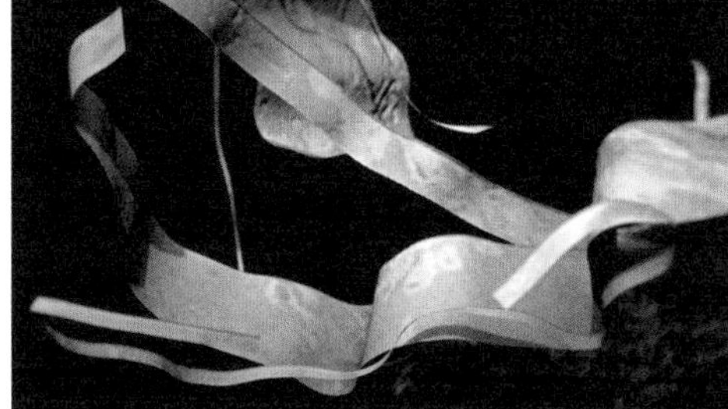

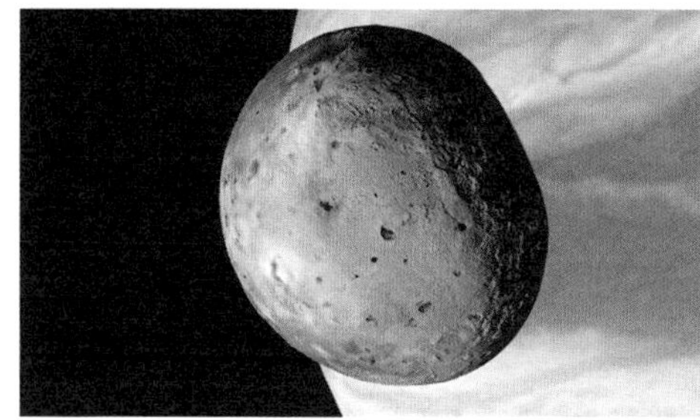

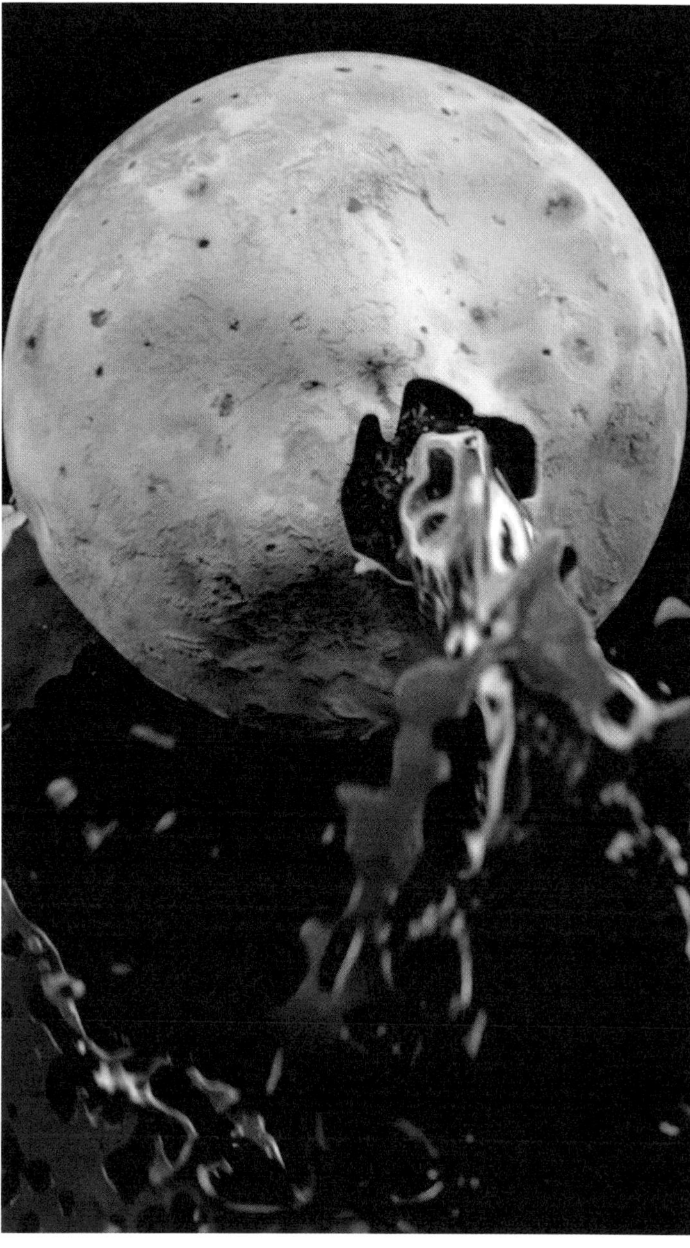

cat. 69
Marie Kølbæk Iversen
IO / I, 2015-

cat. 90 + cat. 92 (t.h.)
Georges Méliès
La Lune à un mètre, 1898
An Astronomer's Dream
Éclipse du Soleil en pleine Lune, 1907
The Eclipse

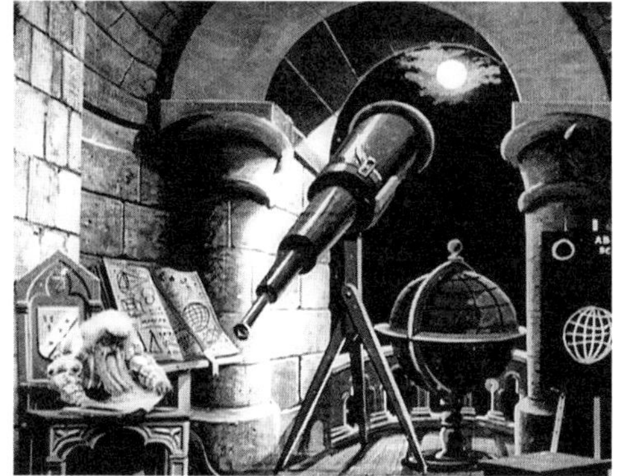

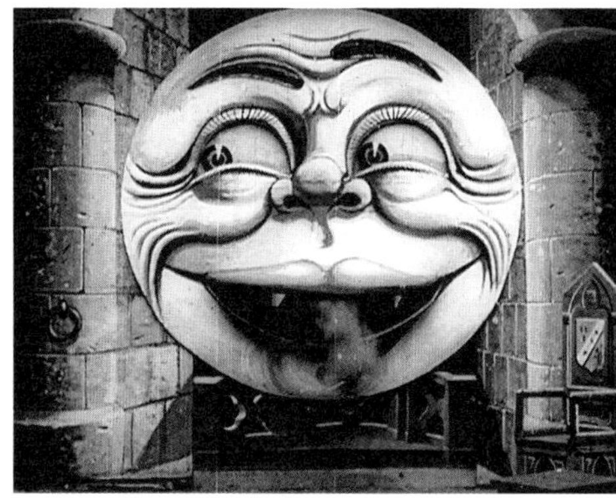

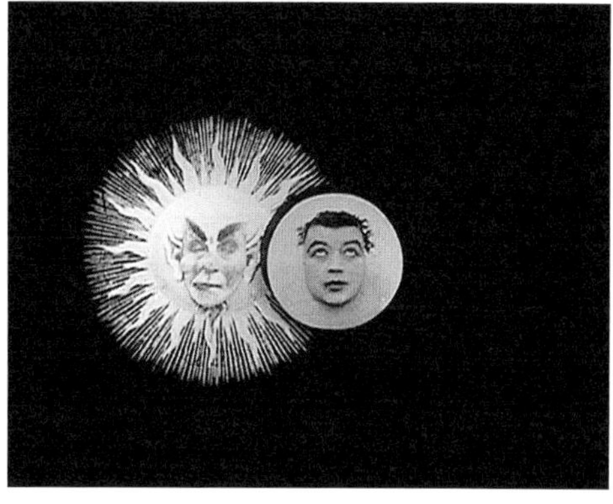

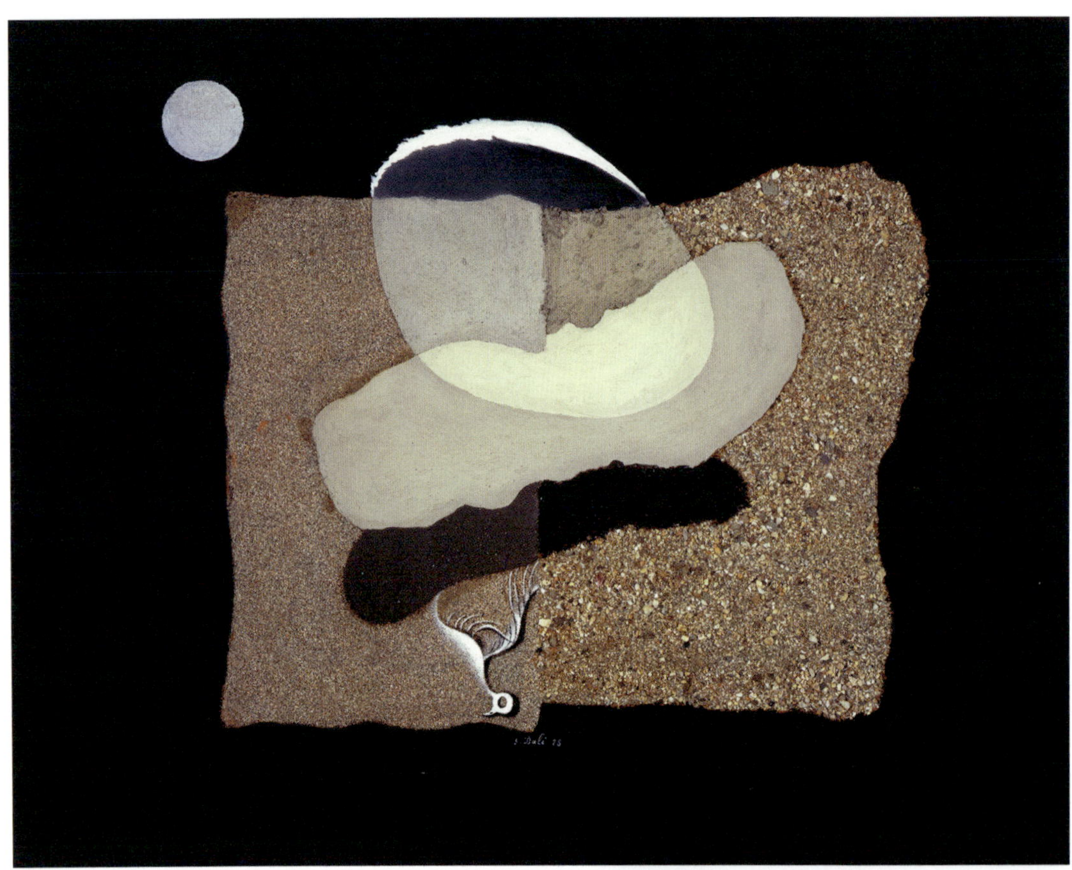

cat. 45
Salvador Dalí
Big Thumb, Beach, Moon and Decaying Bird, 1928

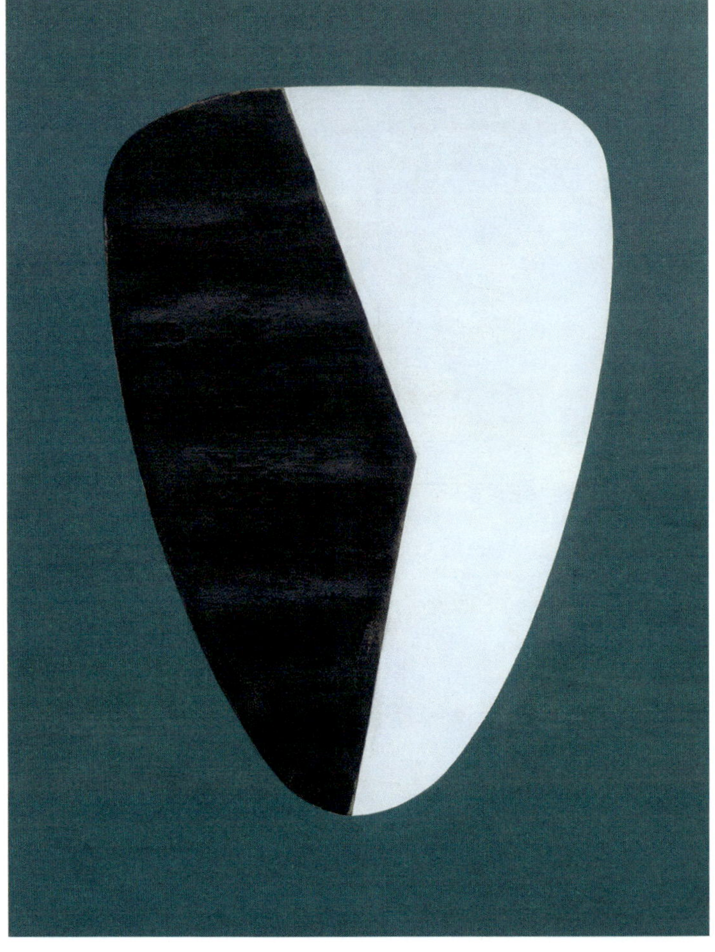

cat. 110
Wolfgang Paalen
Cadran Lunaire, 1935

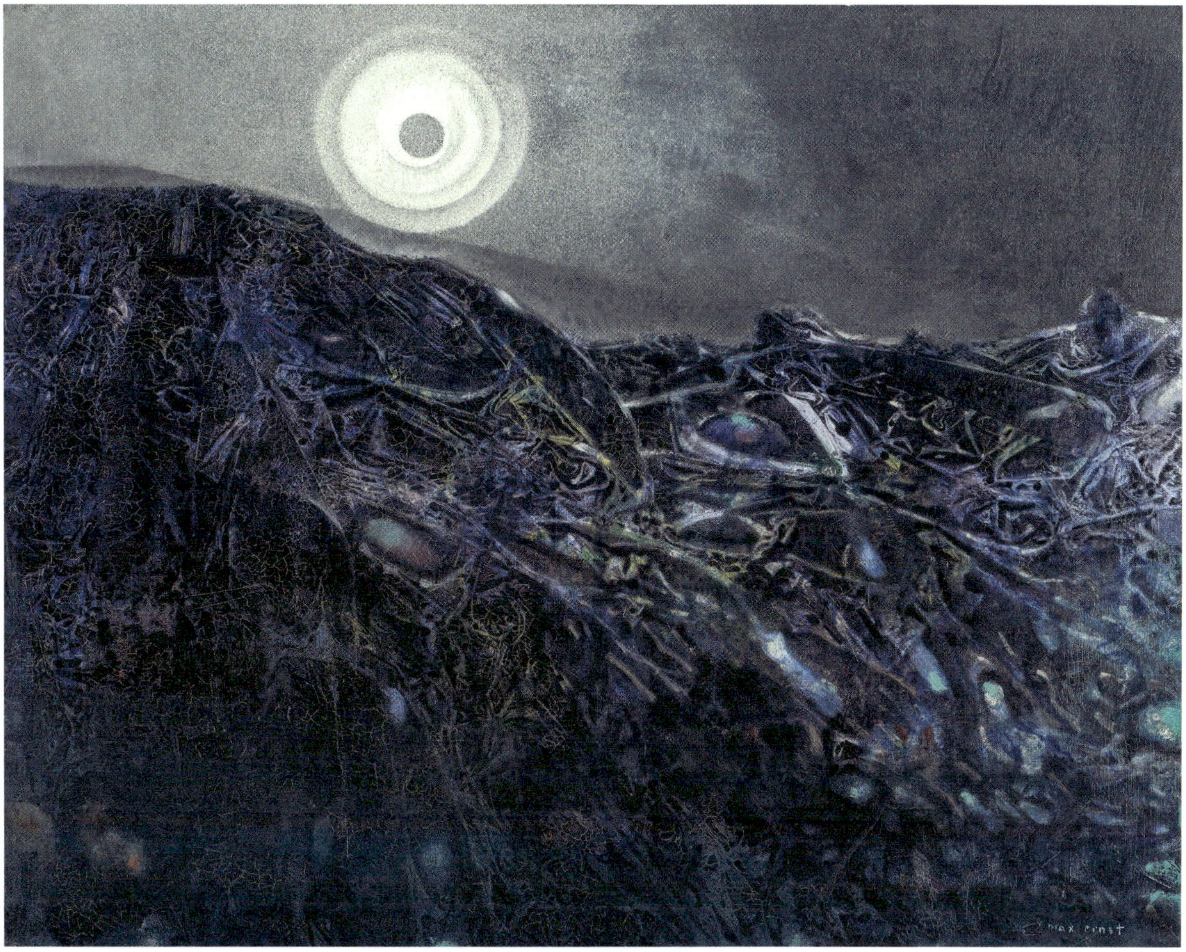

cat. 55
Max Ernst
The Twentieth Century, 1955

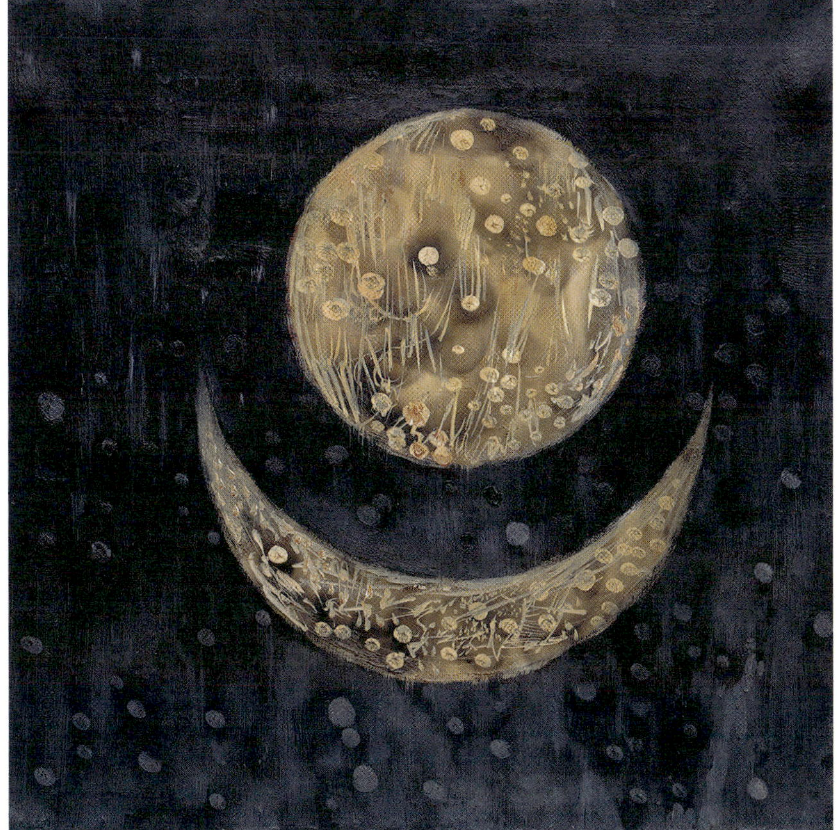

cat. 111
Wolfgang Paalen
Untitled, 1940

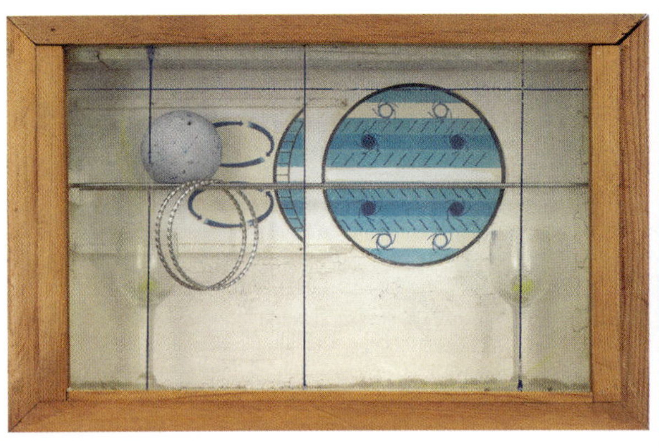

cat. 33
Joseph Cornell
Sun Box, 1960

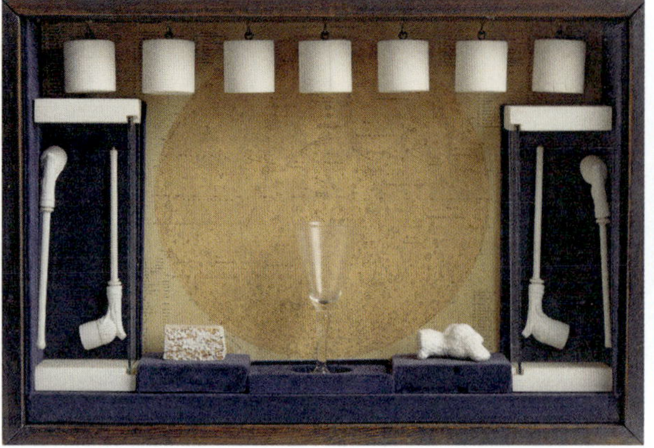

cat. 32
Joseph Cornell
Soap Bubble Set, 1948

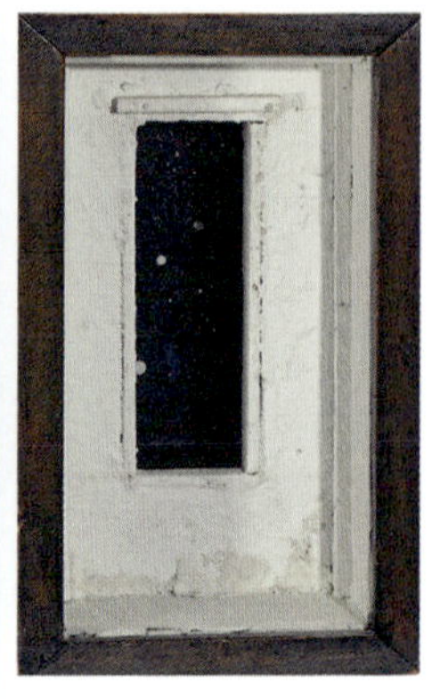

cat. 36
Joseph Cornell
Dovecote, c. 1954

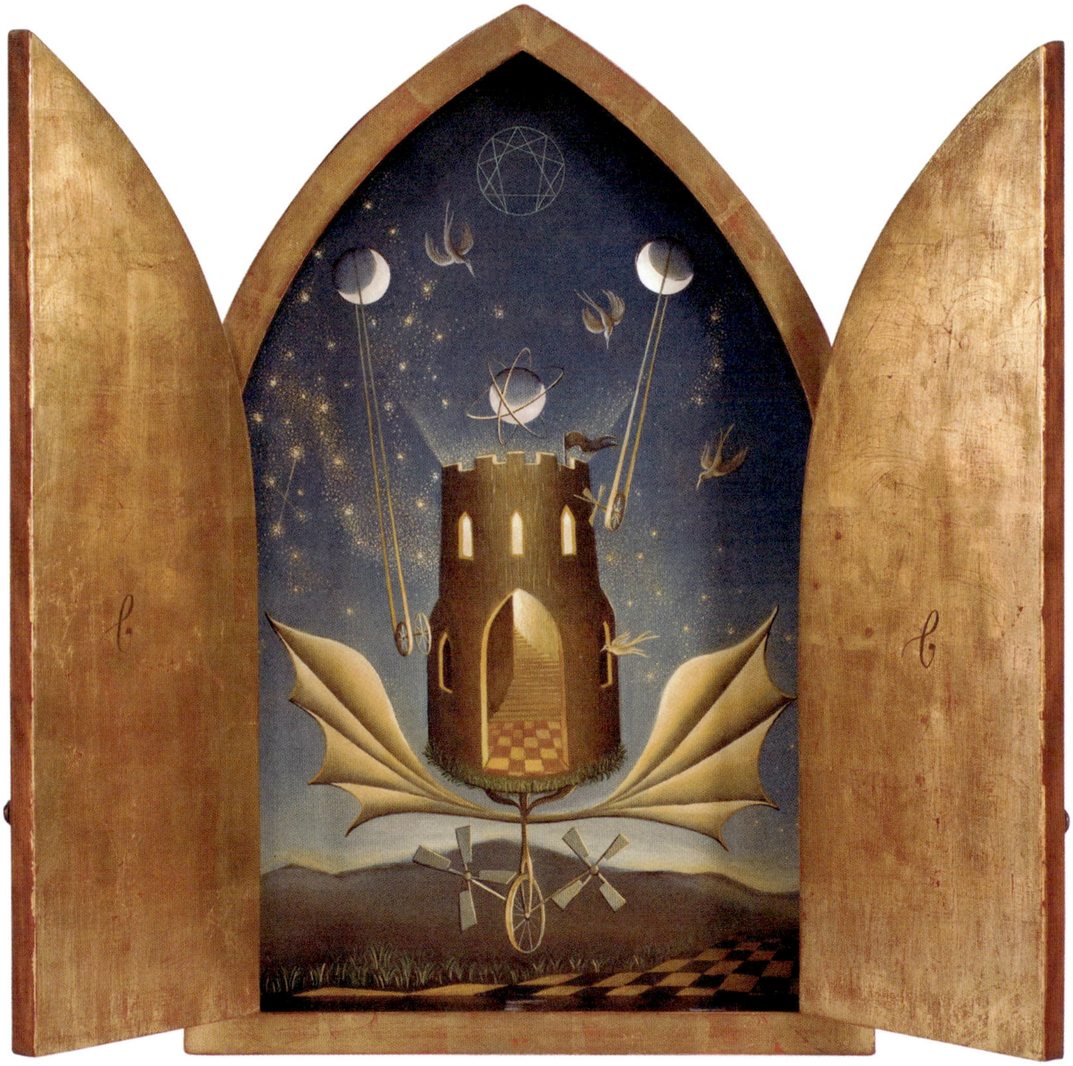

cat. 204
Remedios Varo
Icono, 1945

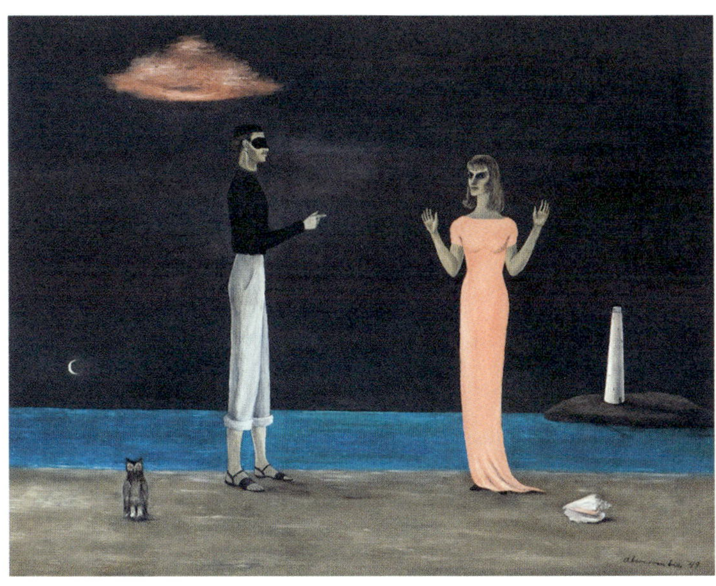

cat. 1
Gertrude Abercrombie
The Courtship, 1949

cat. 205
Remedios Varo
Portrait of Dr. Ignacio Chávez, 1957

cat. 44
Salvador Dalí
Girl with Curls, 1926

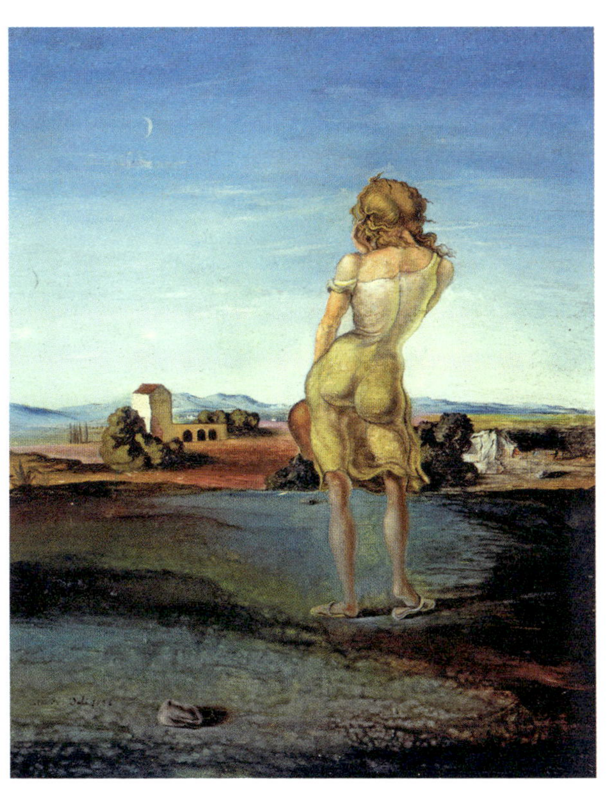

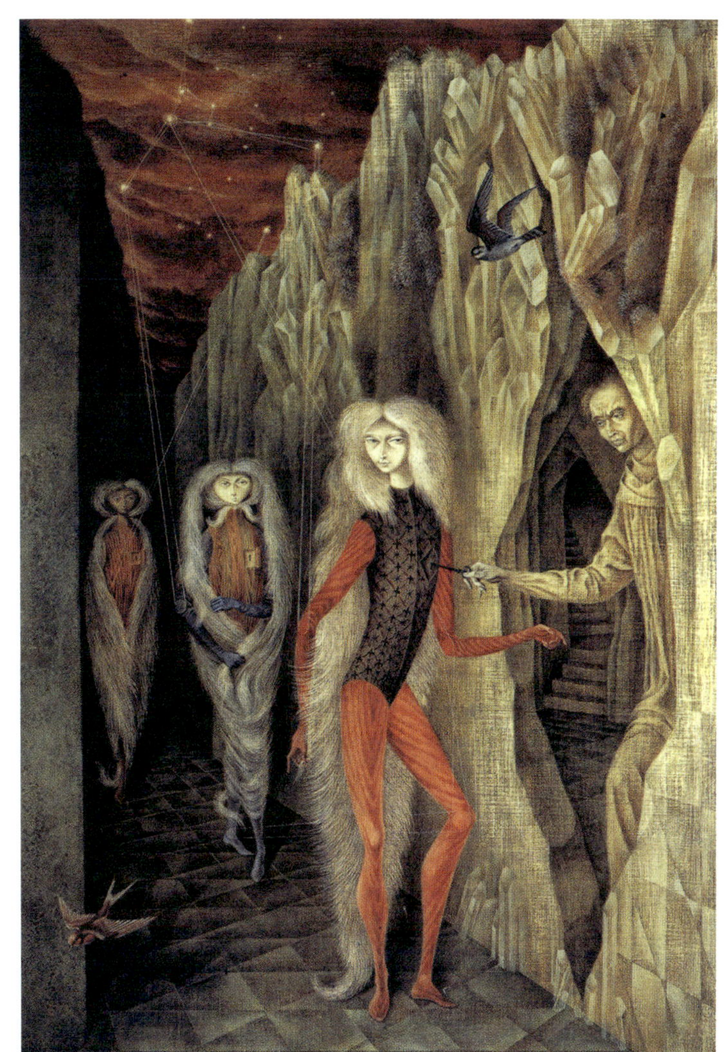

cat. 60
Galileo Galilei
Sidereus Nuncius, 1610
Starry Messenger

Hæc eadem maculã ante ſecundam quadraturam nigrioribus quibuſdam terminis circumuallata conſpicitur; qui tanquam altiſſima montium iuga ex parte Soli auerſa obſcuriores apparent, quà verò Solem reſpiciunt lucidiores extant; cuius oppoſitum in cauitatibus accidit, quarum pars Soli auerſa ſplendens apparet, obſcura verò, ac vmbroſa, quæ ex parte Solis ſita eſt. Imminuta deinde luminoſa ſuperficie, cum primum tota fermè dicta macula tenebris eſt obducta, clariora mõtium dorſa eminenter tenebras ſcandunt. Hanc duplicem apparentiam ſequentes figuræ commoſtrant.

C 2 Vnum

cat. 47
John W. Draper
First known photograph of the Moon, c. 1839-40.
The spots in this photo are caused by mold and water damage on the original daguerreotype, which apparently no longer exists

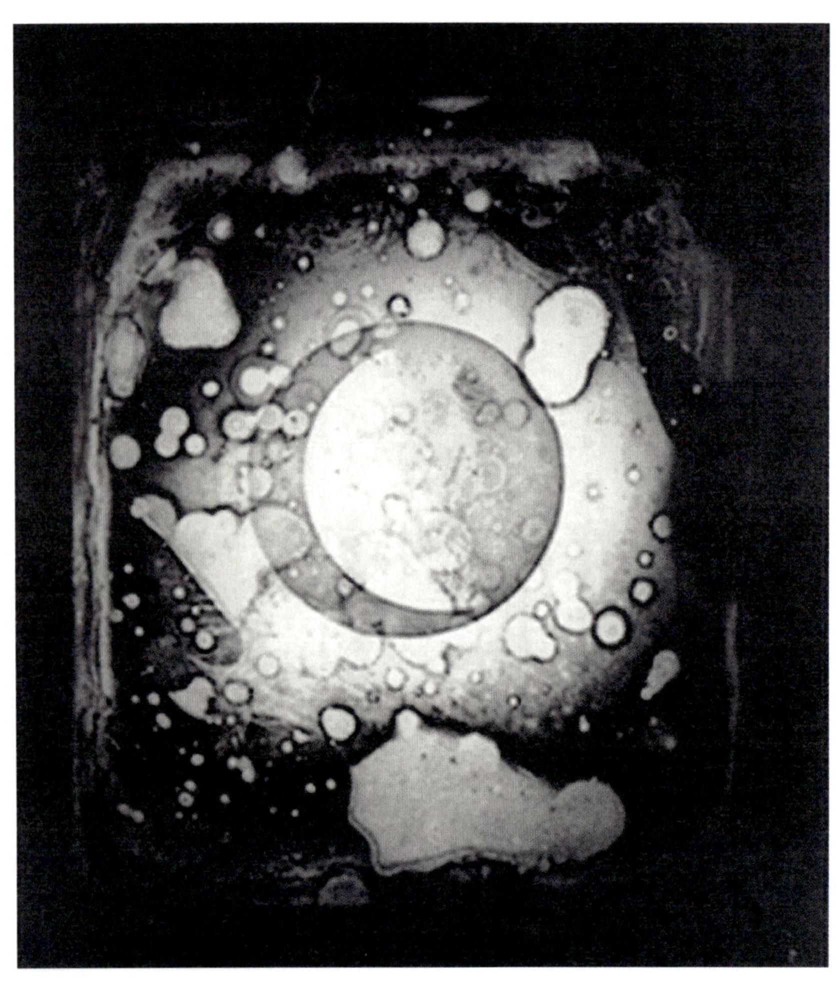

cat. 59
Galileo Galilei
Astronomia, manoscritto cartaceo, autografo, 1609
Watercolour drawing of the phases of the Moon from the autograph manuscript redaction of *Sidereus Nuncius*

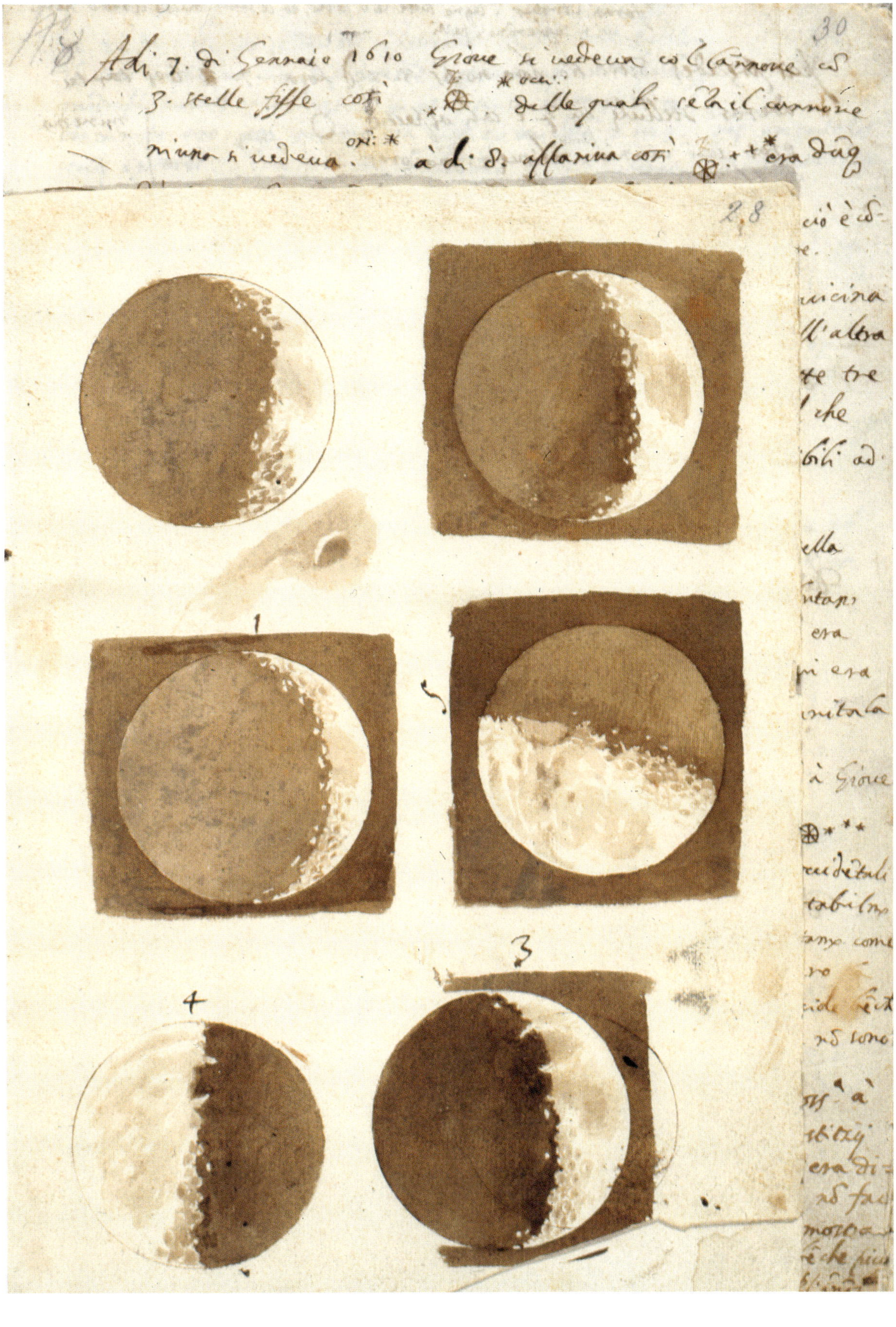

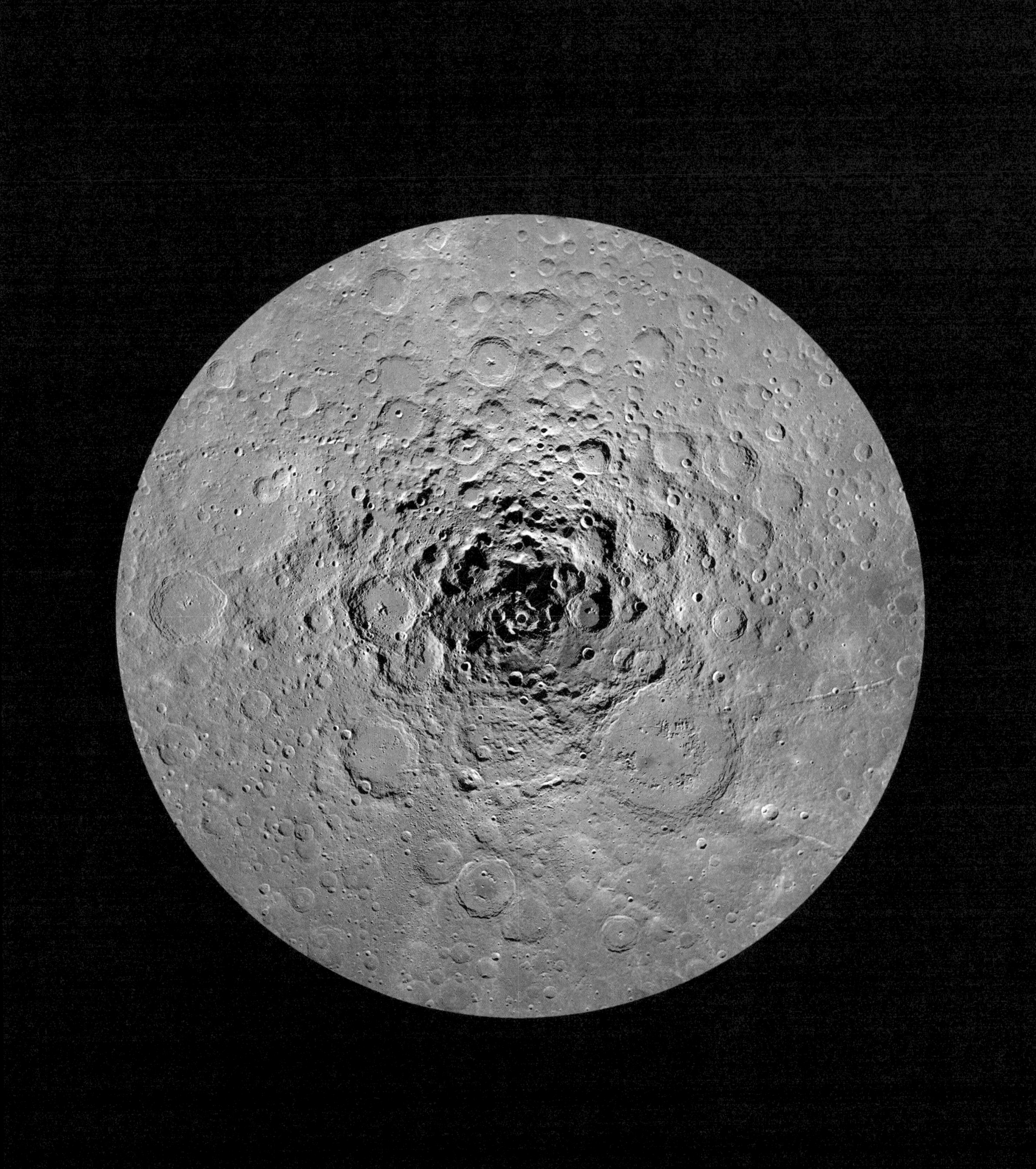

cat. 96
James Nasmyth
Images from the book
The Moon: Considered as a Planet, a World, and a Satellite, 1874

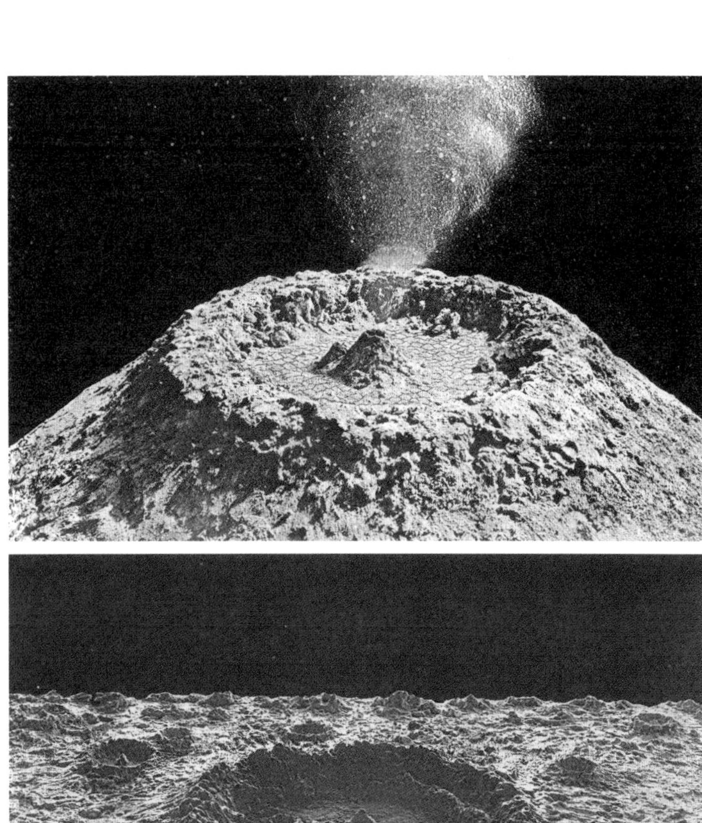

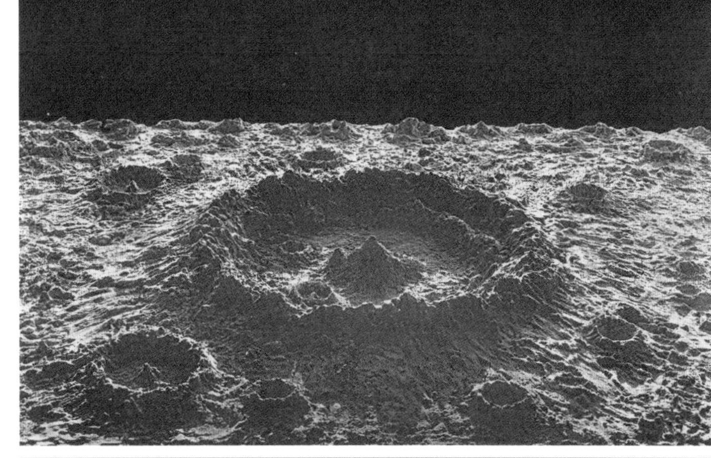

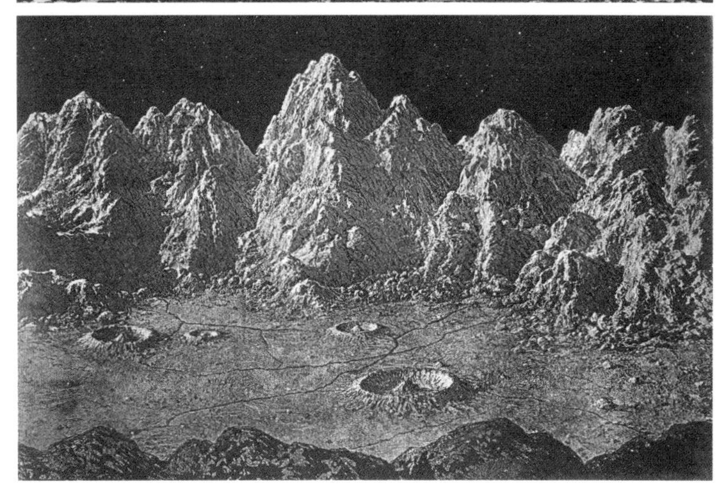

cat. 179
The moon photographed by NASA's Lunar Reconnaissance Orbiter Camera: South Pole

cat. 28
Page from Tycho Brahe's observations of a lunar eclipse from his observatory in the Danish island Hven, 30 December 1590

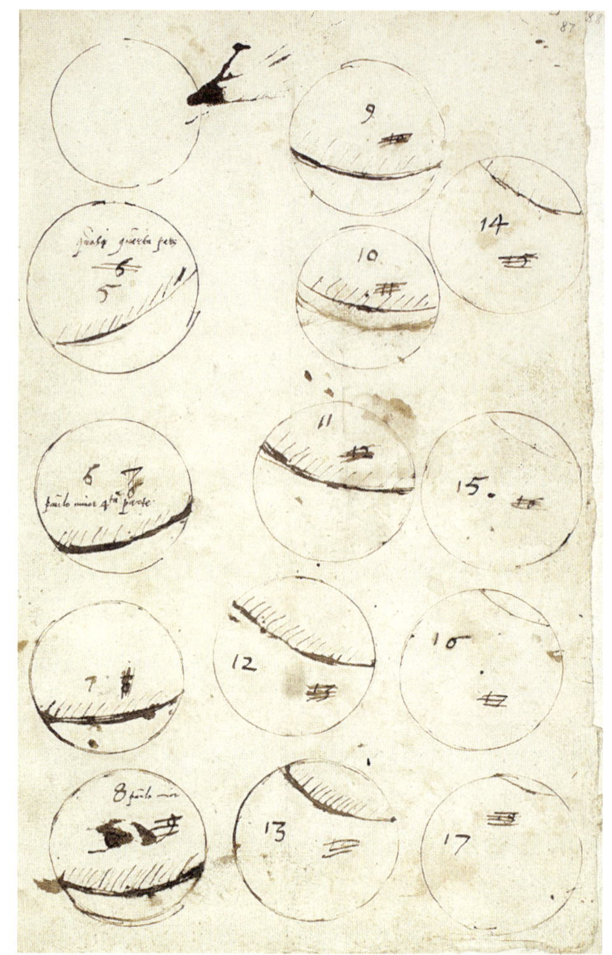

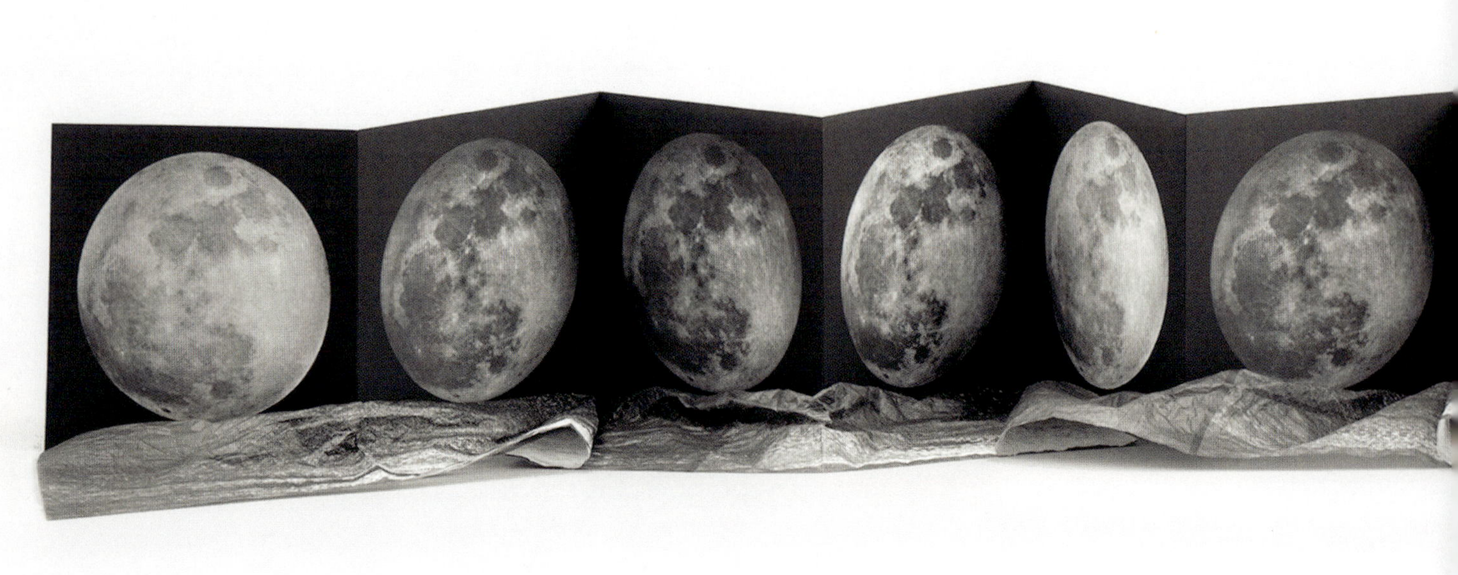

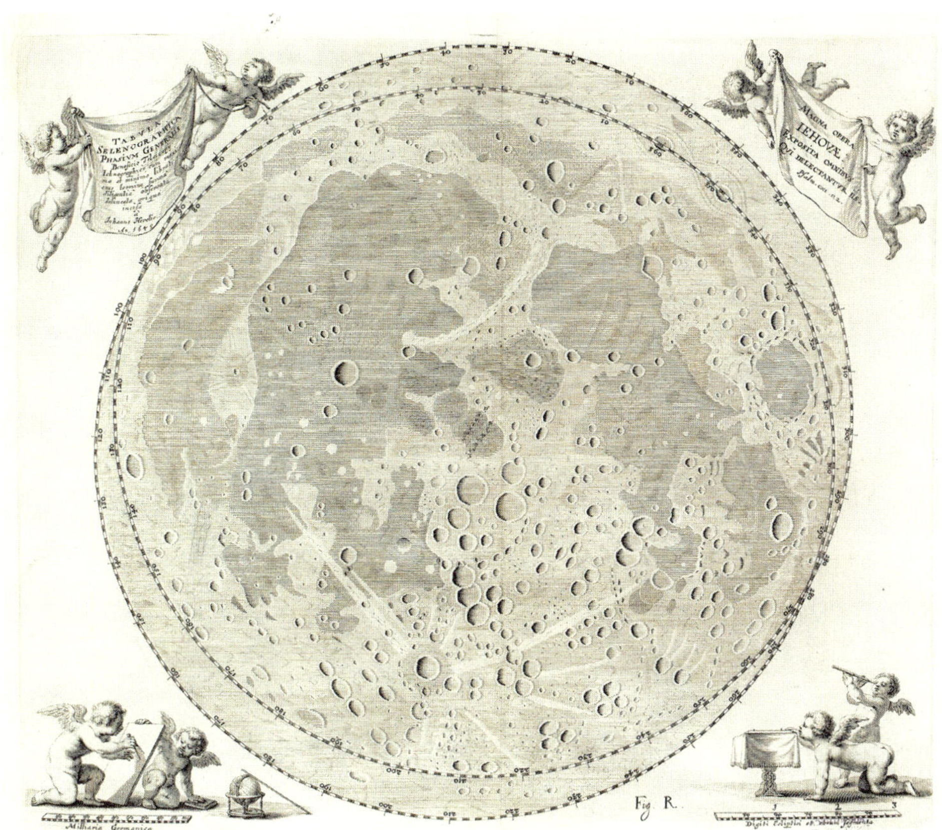

cat. 66
Page from Johannes Hevelius' scientific work *Selenographia* from 1647, the first ever in history to be dedicated solely to the moon

cat. 161
Kiki Smith
Tidal, 1998

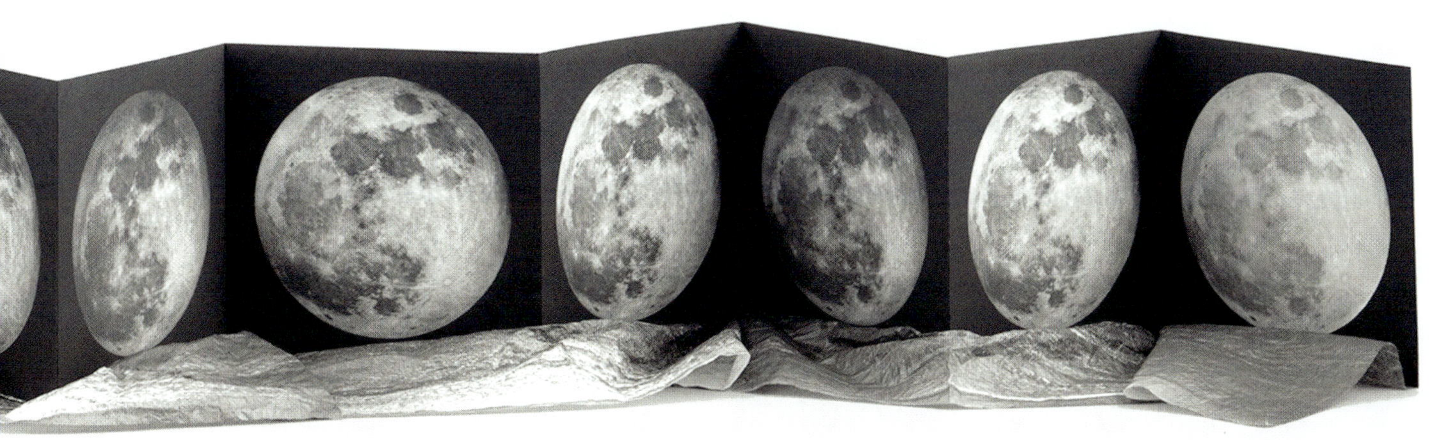

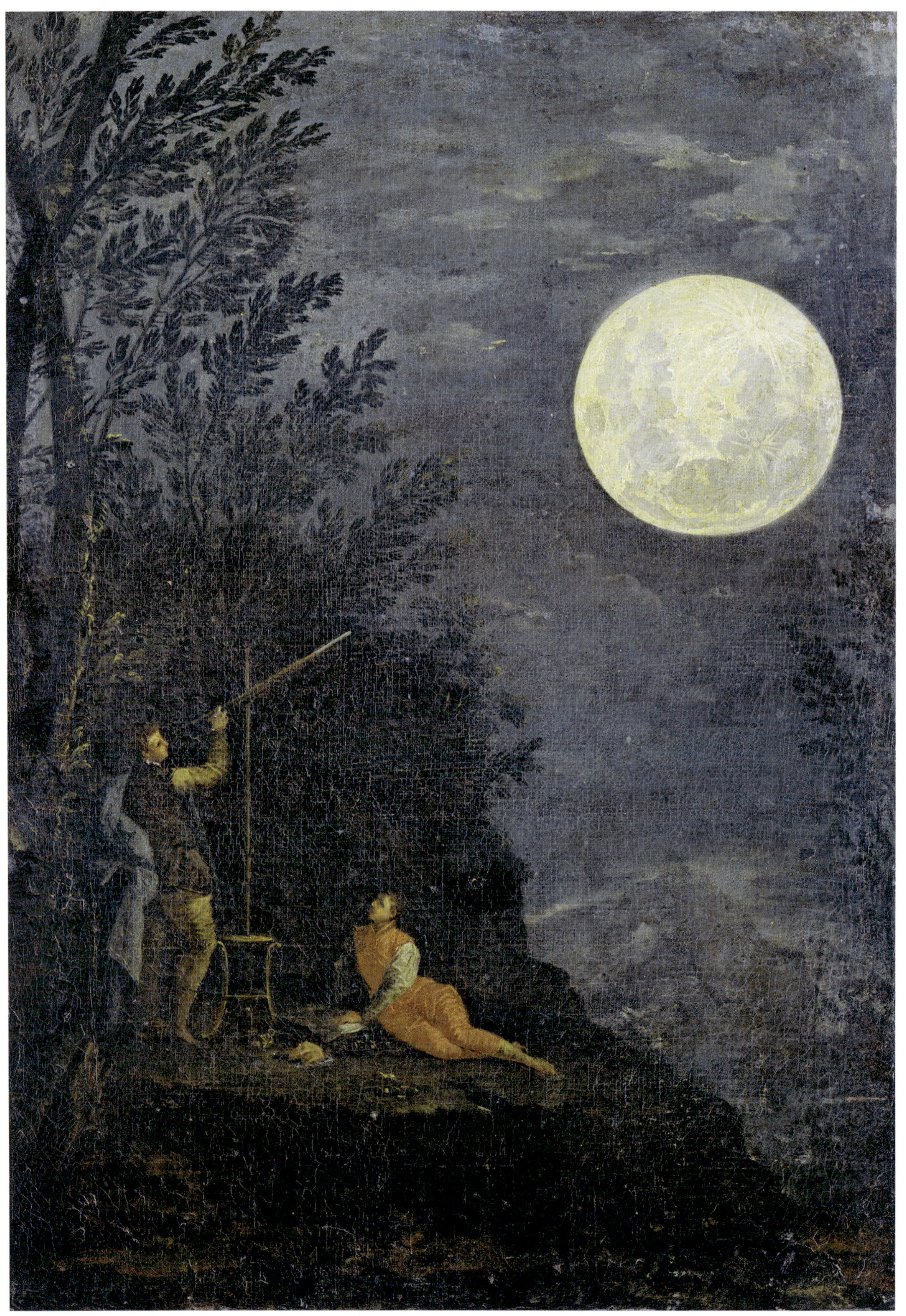

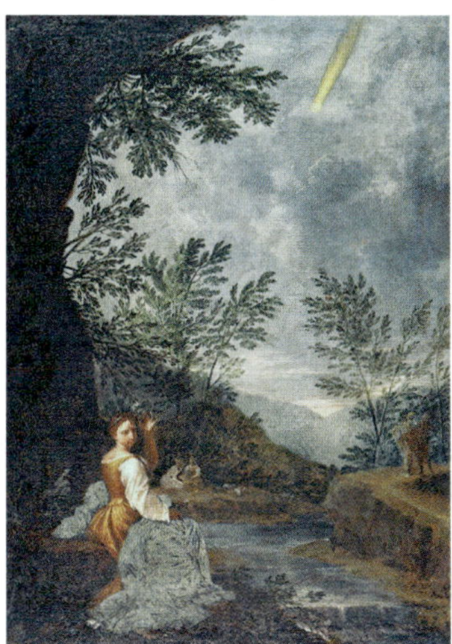
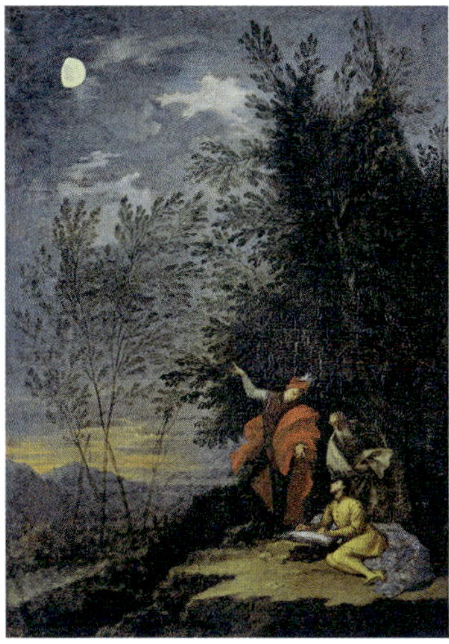
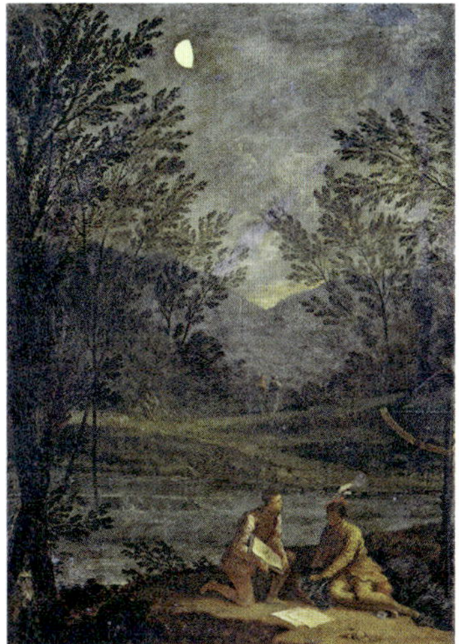
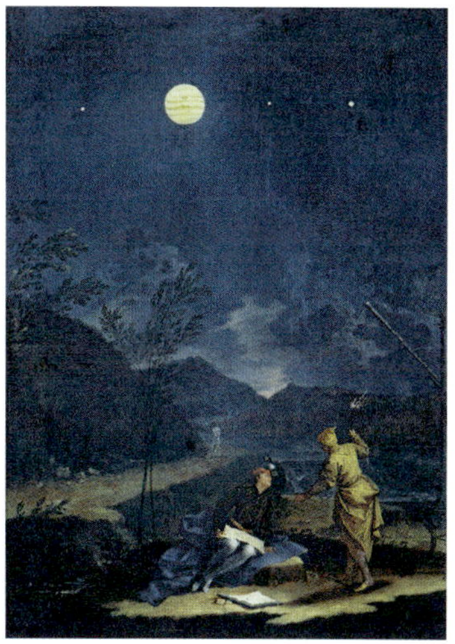

Left:
The Moon

Above:
Comet, Mars,
Mercury, Jupiter

Right:
Venus, Saturn,
the Sun

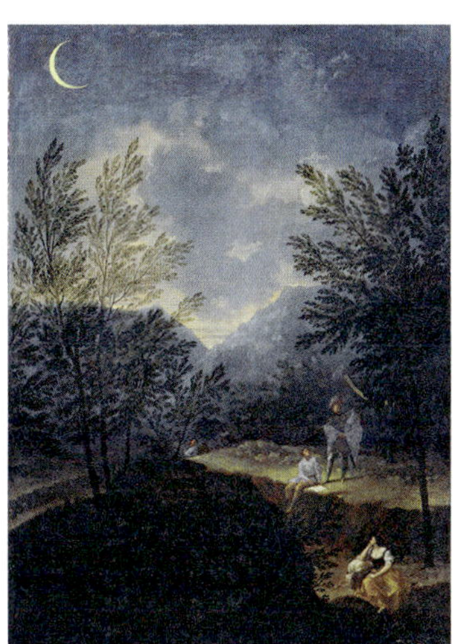
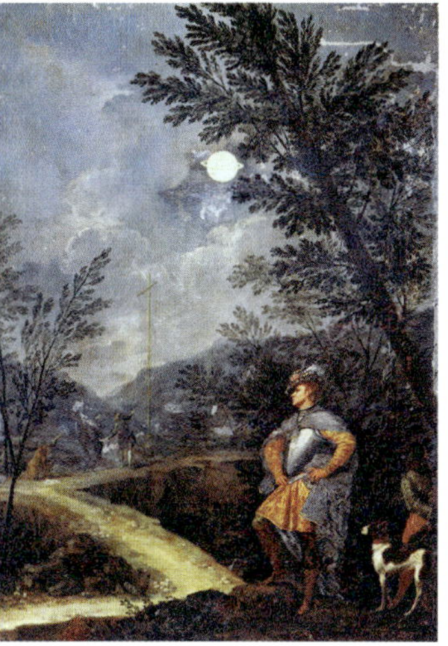
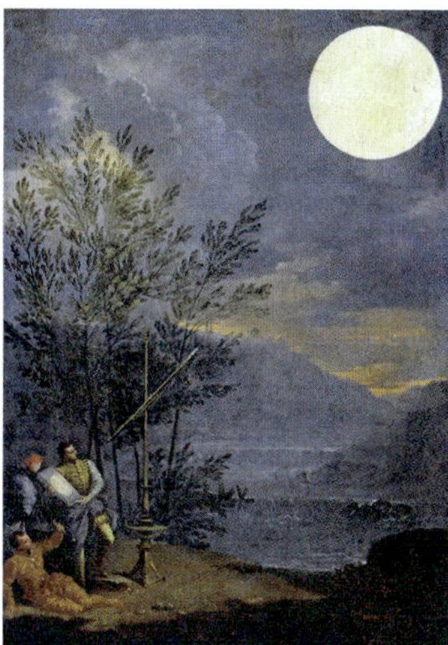

cat. 39
Donato Creti
Le osservazioni astronomiche, 1711
Astronomical Observations

cat. 46
Charles Delagrave's European Celestial Globe, 1878

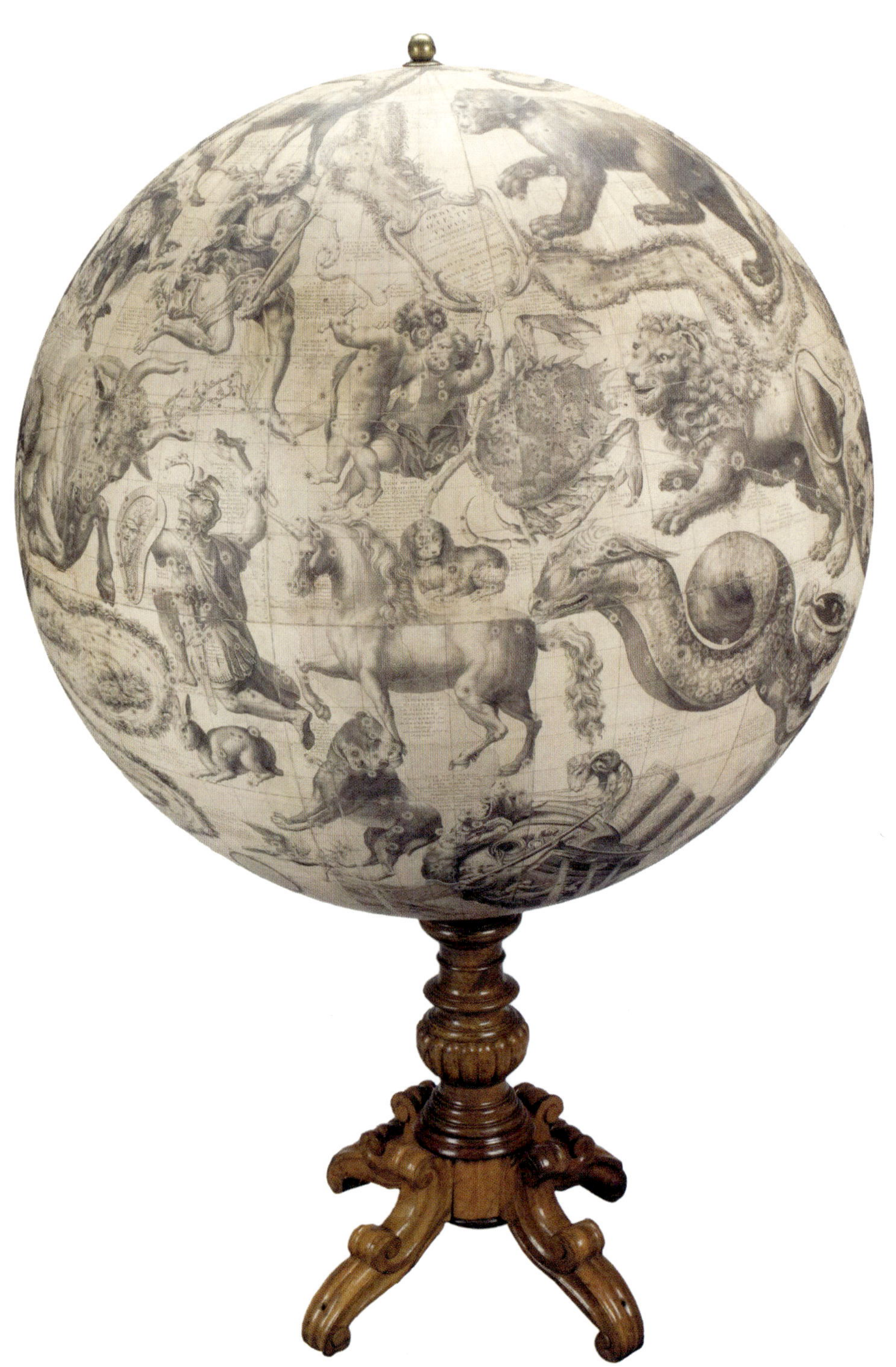

cat. 13
Page from Abd al-Rahman al-Sufi's book of constellations from 964 (this volume 1675), which builds on ancient Greek knowledge of astronomy and became important for European astronomy

cat. 2
Muhammad ibn Ahmad al-Mizzi's Astro Lab Quadrant for measuring the height of the stars at a certain latitude in order to measure time and exact distances for both astronomers, seamen and pilgrims travelling to Mecca, 1329-30

cat. 208
John Adams Whipple & George Phillips Bond's daguerreotype of the Moon in leather case, 1851

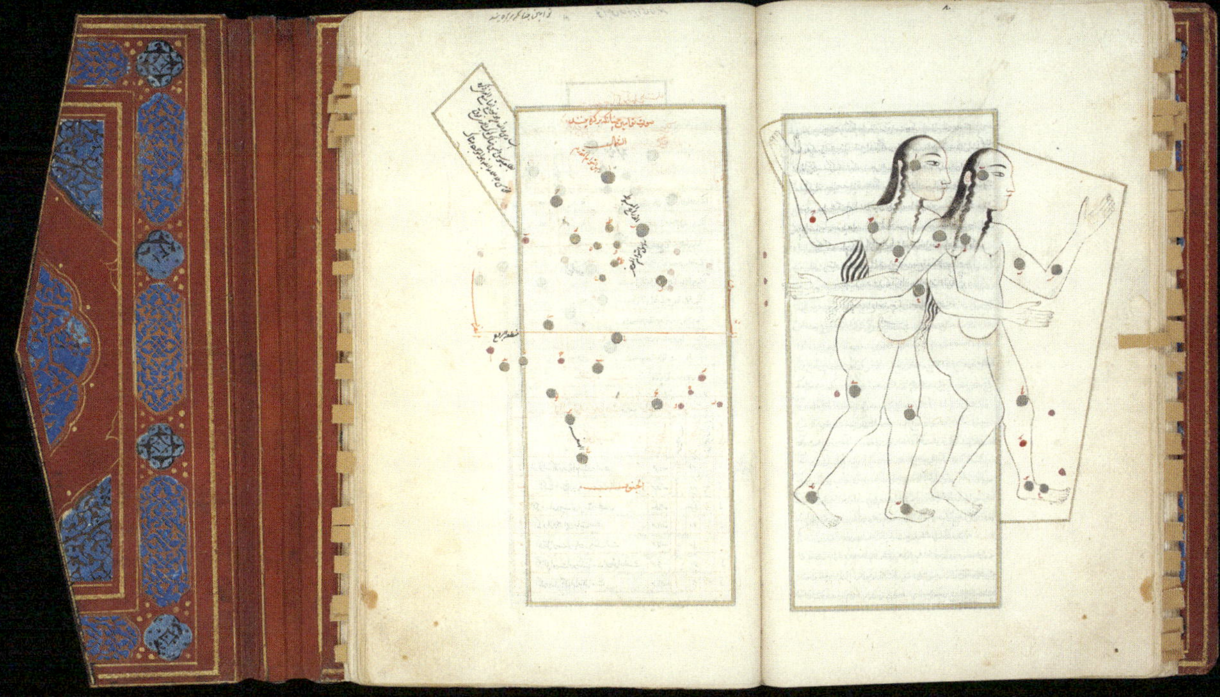

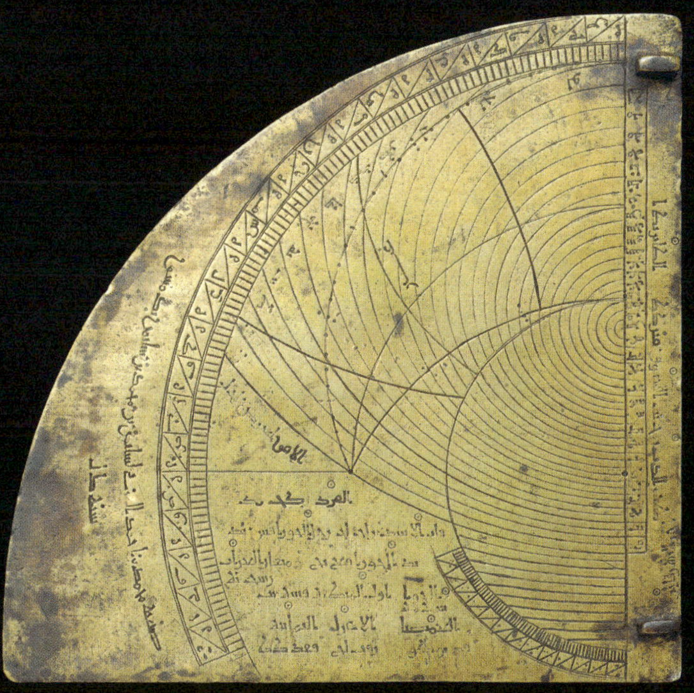

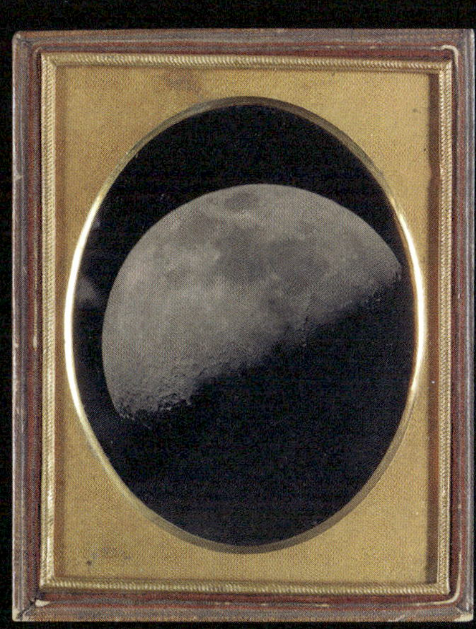

cat. 173
Malena Szlam
Lunar Almanac, 2013

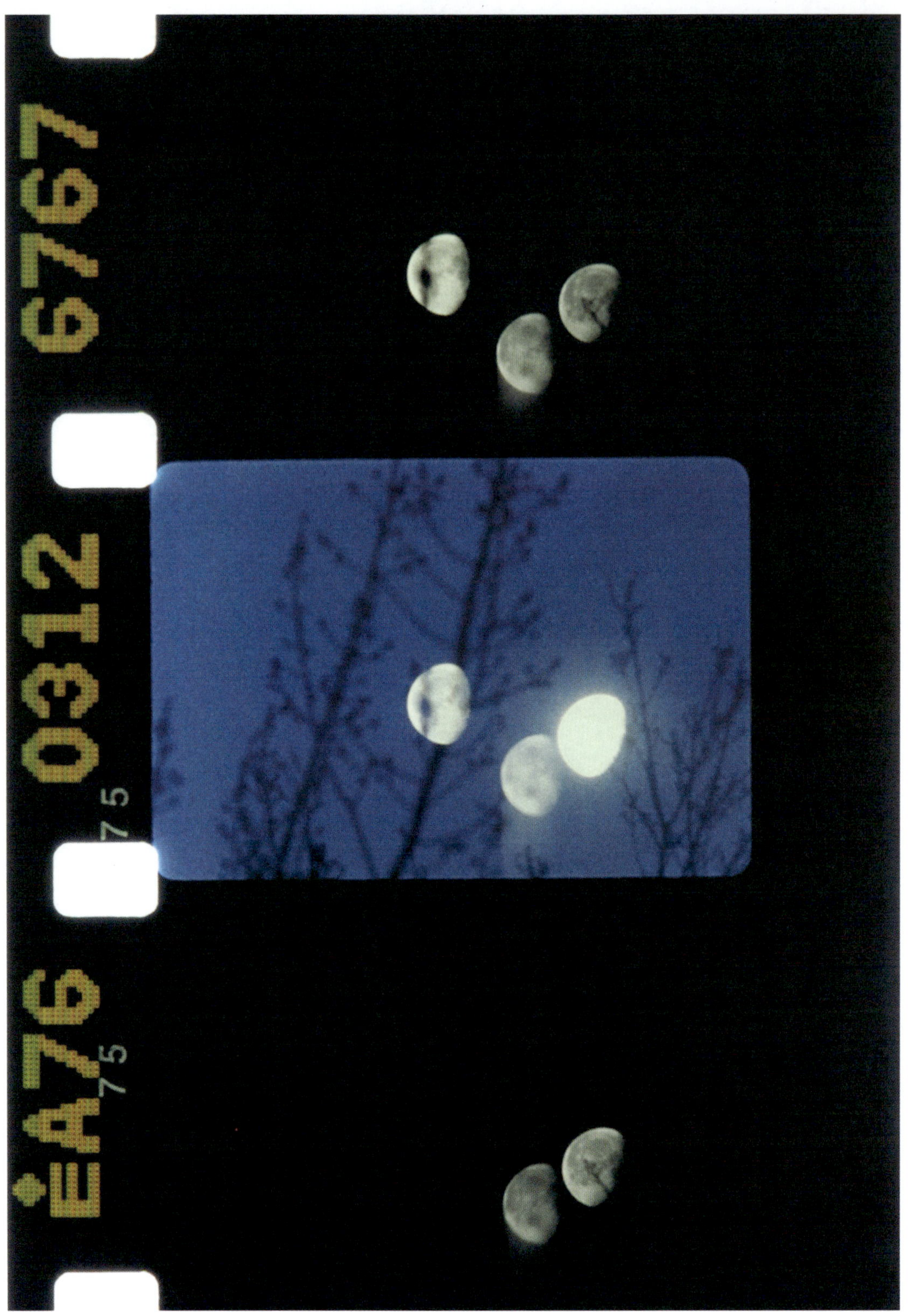

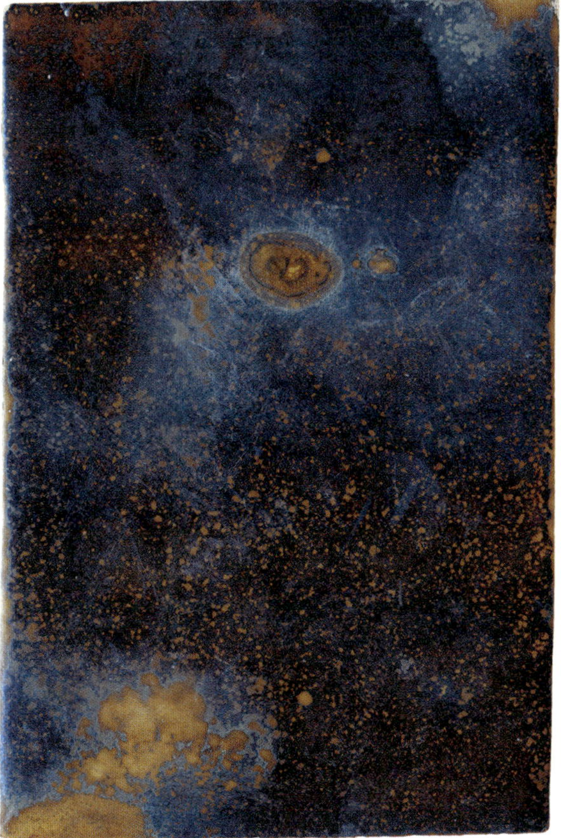

cat. 164
August Strindberg
Celestograph I. Text on the back of the photo: The Full Moon. Without a camera. Exposed for 1 minute and 1½ minutes. Unfixed!, 1893-1894

cat. 169
August Strindberg
Celestograph III. Text on the back: Photograph of the moon without camera or lens. The plate was in the developer for 3/4 hours and exposed to the moon that was very high in the sky. Dornach, early spring 1894

cat. 168
August Strindberg
Celestograph XIII, 1893-94

cat. 75
Yves Klein with the collaboration of Harry Shunk and John Kender
Leap into the Void, 1961

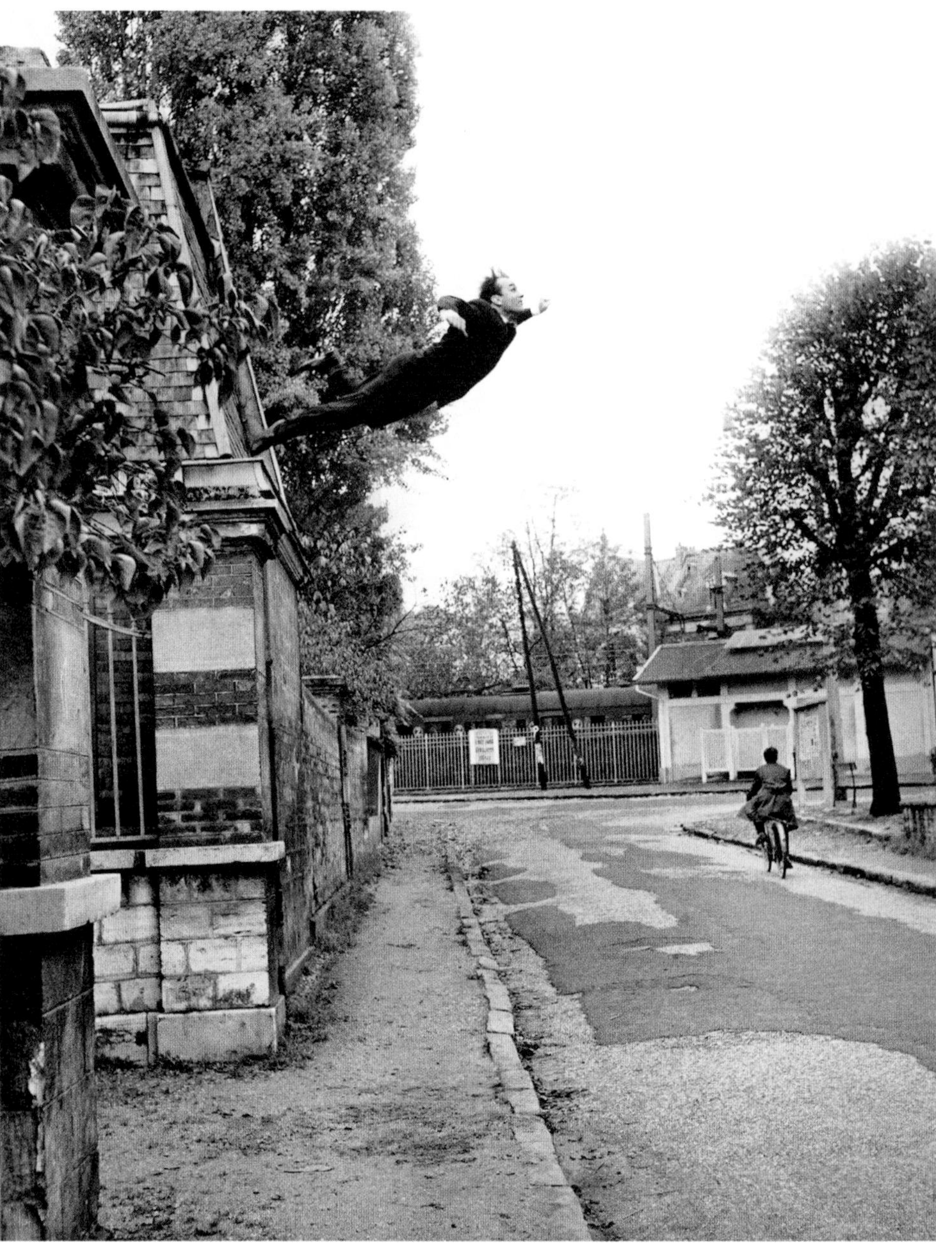

Man and space – space and man – who can draw the dream/reality line …?

– Henry Hopkins
quoted in Robert Rauschenberg, *Stoned Moon Book*, 1970

Stephen Petersen is an American independent scholar and author of the book *Space-Age Aesthetics: Lucio Fontana, Yves Klein, and the Postwar European Avant-Garde* (Pennsylvania State University Press, 2009).

Astronautic Theater: Space Flights and Lunar Expeditions in 1960s Art

By Stephen Petersen

As part of the burgeoning Happenings scene in New York at the start of the 1960s, Robert Whitman produced an ambitious, multimedia performance that he called *American Moon* (p. 86). Held over several evenings at the Reuben Gallery in November and December of 1960, Whitman's dynamic event featured an elaborately devised stage environment, dance performers, otherworldly sculptural props, film projections, lighting effects, inflatable plastic structures, and audience participation. It culminated with the revelation of a figure (the artist Lucas Samaras) poised on a swing hanging above the spectators, like an astronaut floating overhead. Whitman had originally planned to have multiple figures "suspended and pulled across the space." He envisioned a space "filled with bodies shooting around in all directions ... hung from the top by huge elastics ..."[1] Although this "gravity-defying event" proved impossible to realize, it remains a tantalizing vision of space flight as artistic liberation at the dawn of the 1960s.

The idea of voyaging to the moon loomed large in postwar culture. In the wake of World War II technological developments, it was no longer a far-fetched dream but rather an anticipated reality. As early as 1946, the *Illustrated London News* ran an article entitled "By Space Rocket to the Moon? Problems, which Prevent Man from Journeying in Outer Space, Explained." The George H. Pal film *Destination Moon* (1950) was conceived, in the director's words, not as fantasy but as a "documentary of the near future."[2] In the context of the Cold War "race for space," John F. Kennedy called in 1961 for the United States to commit itself to "achieving the goal, before this decade is out, of landing a man on the Moon and returning him safely to earth."

In their own ways, artists sought to participate in this space enterprise. The Japanese American sculptor and designer Isamu Noguchi produced an unusual series of "lunar" reliefs in the late 1940s that culminated in a group of environmental interiors including *Lunar Voyage* (1948), designed for a stairway in the ocean liner the S.S. Argentina. These works combined biomorphic abstract forms with colored lights to suggest an otherworldly atmosphere. In a similar spirit, Italian artist Lucio Fontana unveiled his first *Spatial Environment* at the Galeria del Naviglio in Milan in 1949. Suspended abstract forms lit by ultraviolet rays appeared to hang as if free of gravity in the darkened space of the gallery. With this work, wrote a critic at the time, "Fontana has touched the Moon."[3]

The prospect of a manned moon mission was publicized increasingly throughout the 1950s. In 1955 the Walt Disney television program titled "Disneyland" featured two space-themed specials that combined science and entertainment. The first, *Man in Space* (p. 91), based on Heinz Haber's book of that name, explored the physical and psychological effects of the zero-gravity environment of outer space. The next, *Man and the Moon*, detailed an imaginary rocket-flight around the moon, with a lengthy live action simulation. Not coincidentally, Disneyland in California had opened its "Rocket to the Moon" ride that same year, offering a simulated lunar voyage to eager crowds.

That year as well, British artist Richard Hamilton realized a large-scale installation entitled *Man, Machine & Motion* (p. 87) at the Hatton Gallery in Newcastle-upon-Tyne. A walk-through grid displayed enlarged panels made from scientific illustrations depicting different types of machine-assisted human travel (underwater, on land, in the air, and, lastly, in space). Overhead, panels displayed images from popular science and science fiction of space-suited astronauts floating though the black void of outer space, along with a scene from the 1950 space epic *Destination Moon* showing astronauts standing on the moon.

The image of the astronaut would play a central role in Hamilton's series of paintings and drawings entitled *Towards a definitive statement on the coming trends in men's wear and accessories* (1962-63), which sought "to represent our mid-century myths, dreams and exploits."[4] The mass cultural heroes featured in Hamilton's series are at once celebrated and deconstructed as ideals of masculinity. The first painting in the series presents the visage of John F. Kennedy in a space-helmet, along with knobs, scan-lines, and other references to the technologies of space flight and mass media. The work's subtitle, drawn from Kennedy's 1961 inaugural speech, is "Together Let Us Explore the Stars." The topic of the picture, said Hamilton, was "man in a technological environment."[5]

French artist Yves Klein approached the iconography of human space exploration from a different perspective, emphasizing mind-powered levitation over technology-assisted aviation: "it will not be with rockets, sputniks, or missiles, that man will realize the conquest of space, because in that way, he would remain always a tourist in space, but it is through inhabiting it in sensibility."[6] In 1960 Klein made headlines for himself (literally) with his *Leap into the Void*, a seamless photomontage depicting the artist leaping upward from a second-story rooftop, wearing eveningwear and appearing to soar under his own power into the sky. Klein published the image in his *Dimanche* (the newspaper of a single day) with the headline, "Un Homme dans l'Espace!": a man in space! Klein, so to speak, got the jump on Soviet Cosmonaut Yuri Gagarin, whose orbital flight a year later would inaugurate the era of manned space flight.

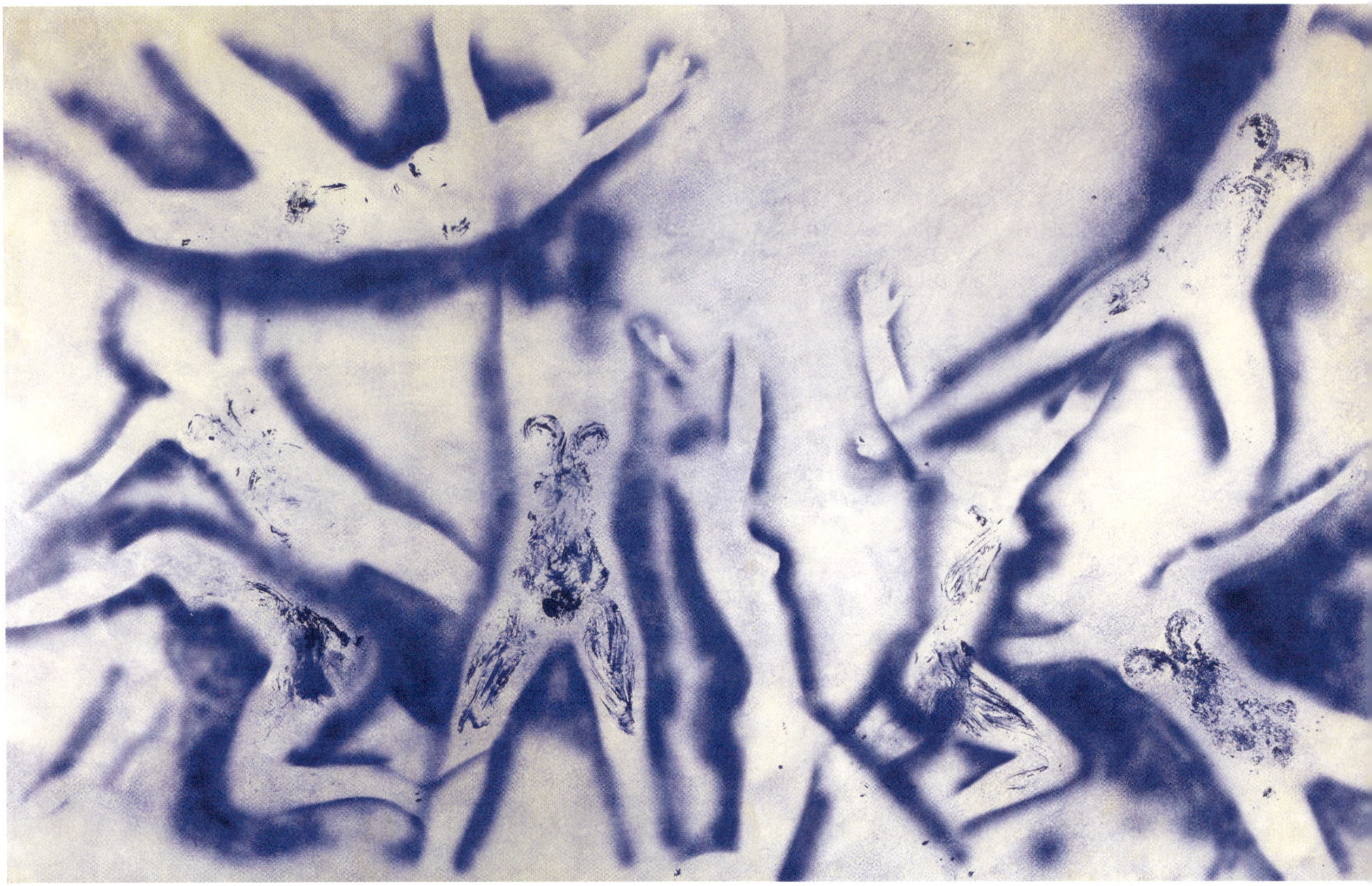

Yves Klein
People Begin to Fly (ANT 96), 1961
The Menil Collection, Houston

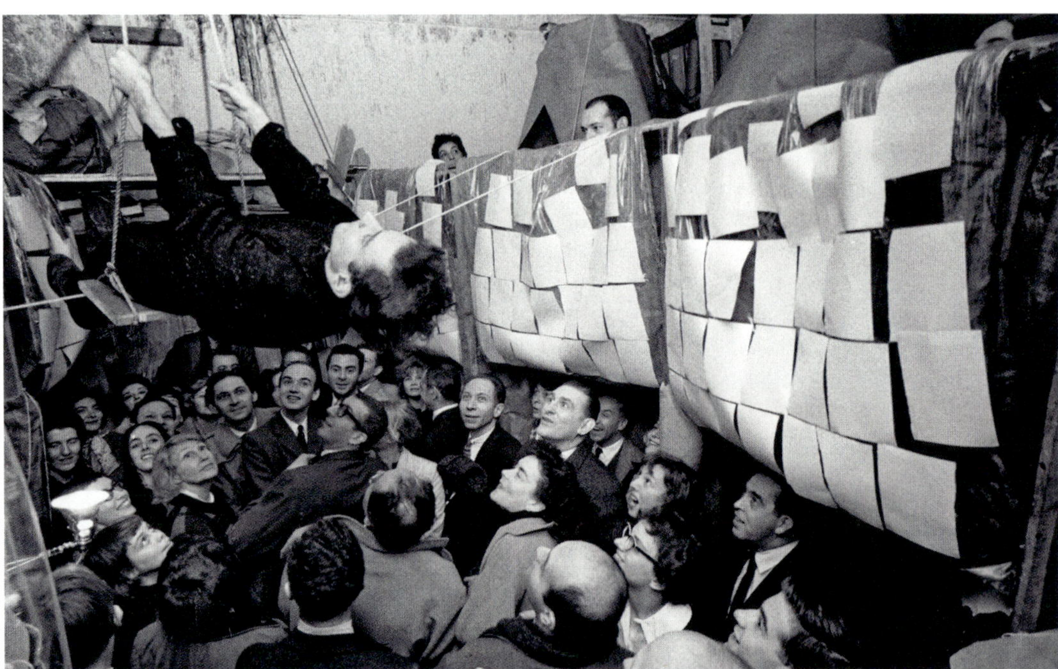

Robert Whitman
American Moon, 1960
Performance

In 1961, Klein produced his large-scale "anthropometry" painting entitled *People Begin To Fly*, featuring life-size imprinted nude female figures assuming various postures of flight, symbolizing for Klein total physical and spiritual freedom. That year, Klein also began his *Planetary Reliefs* (pp. 105-107), terrestrial and extraterrestrial topographies coated in luminous pigment. When Klein exhibited these works for the first time in Milan in 1961, two rose-colored reliefs of the moon's surface were, in critic Pierre Restany's words, "the sensation of the exhibit."[7] Despite their strangeness, Klein's extraterrestrial reliefs were, said gallery-owner Guido le Noci, "more 'real' than any astronomical observation."[8]

For the Austrian artist Kiki Kogelnik, active in New York in the mid-1960s, space flight likewise offered a complex mythology and an imaginary liberation, as well as a source of humor, evident in her schematic pop renditions of rockets, robots, planets, and astronauts. In the erotically charged and optically vibrant *Fly Me to the Moon* (1963, p. 7) and *Outer Space* (1964), life-sized silhouettes of the female body with simplified contours and discs for heads ascend in a flattened pictorial space. The figures in Kogelnik's paintings can be seen as stand-ins for the artist – inherently a subversive take on the astronaut as a masculine ideal. Kogelnik explored the choreography of the body in space using herself as performer, producing an experimental film, *Untitled (Floating)* (1963). According to Ciara Moloney, Kogelnik "performed a series of movements in the studio before an 8mm camera turned upside down to mimic the weightlessness of space travel."[9] That same year, the female Cosmonaut Valentina Tereshkova became, at age 26, the first, and to this date the youngest, woman to fly in space. Tereshkova orbited the earth some 48 times over three days; meanwhile, nearly 20 years would pass before another woman went into space.

Kogelnik engaged with the space program perhaps most directly with her 1969 performance *Mondlandung!,* held at the Galerie Nacht St. Stephan in Vienna the morning of July 21 to coincide exactly with the Apollo lunar landing and moonwalk. As the televised coverage of the moonlanding appeared on a monitor in the gallery, she completed a silkscreen depicting the disc of the moon, overlaid with the names of the three American astronauts and the words of one of Armstrong's first pronouncements, "I Can See My Footprints" (p. 89). The screenprints were available at a special price for those in attendance; meanwhile, a breakfast buffet and "Moon strudel" were served. Kogelnik was attempting to respond in real time to the mediated event taking place – reporting Armstrong's words as she heard them. Her act of identification as an artist making a print, with the astronaut

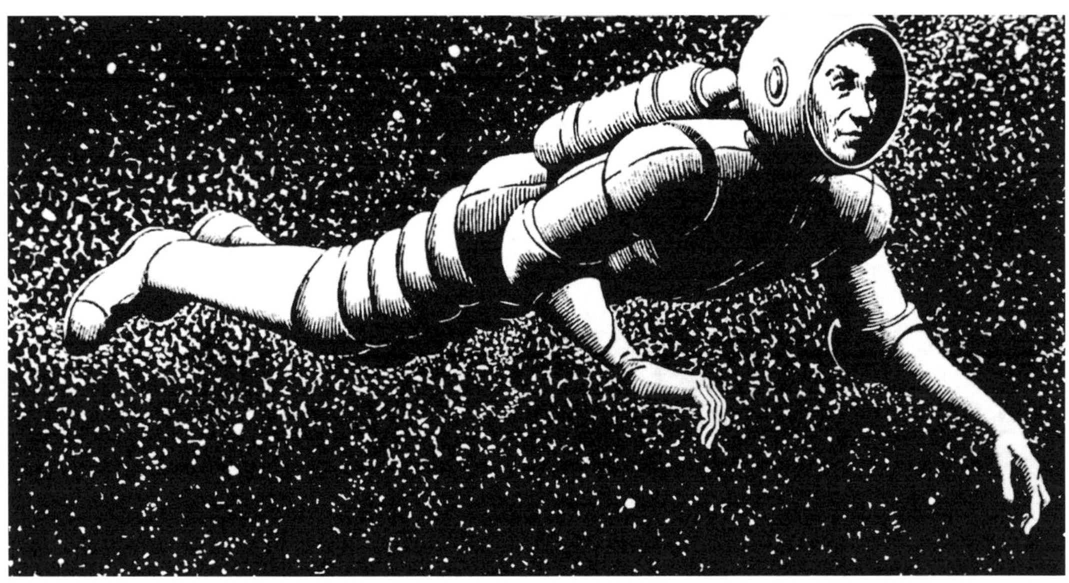

Richard Hamilton
Panel from the exhibition *Man, Machine & Motion*, 1955

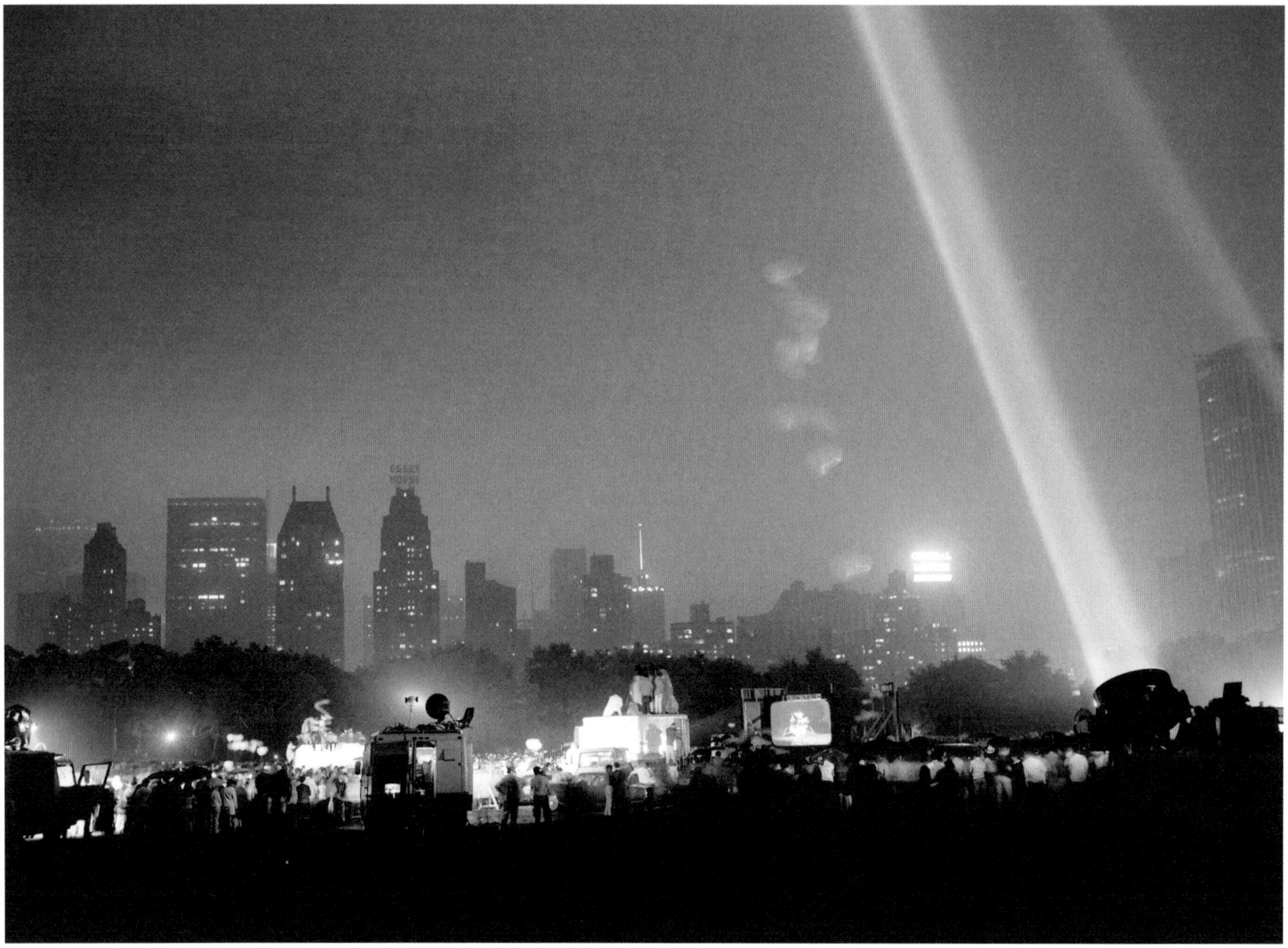

'Moon-In' in Central Park, 20 July 1969
NYC Parks Photo Archive

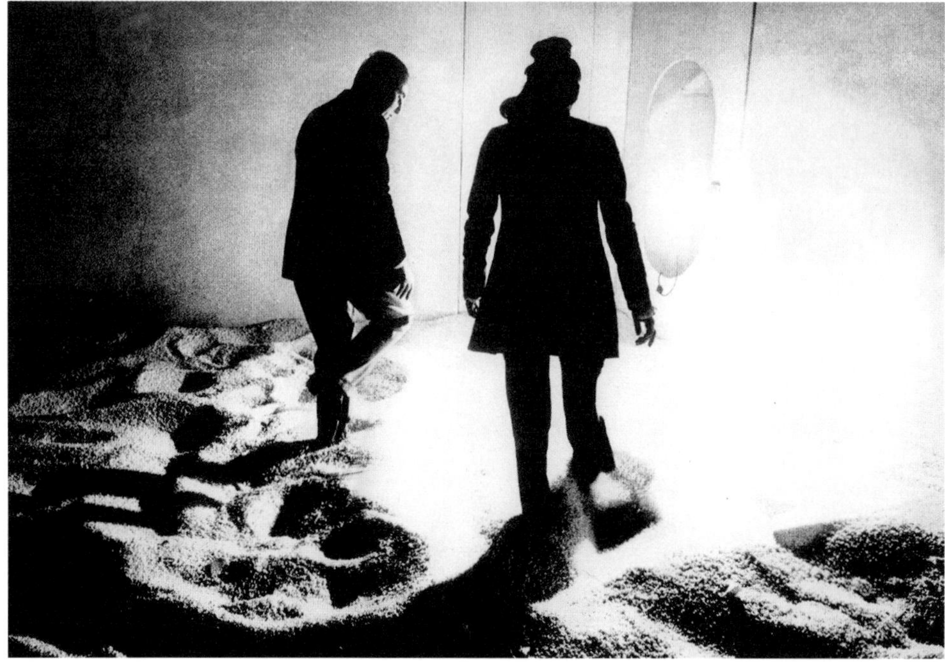

Fabio Mauri
Luna, 1968
Courtesy The Estate of Fabio Mauri and Hauser & Wirth

Kiki Kogelnik
I can see my Footprints, 1969,
Courtesy of Kiki Kogelnik Foundation

seeing his footprints, proclaims both her closeness to and, in other ways, her distance from the event itself. Adding further layers of mediation, she filmed and photographed her "Moonhappening" as well.

A happening of a different kind took place in New York City to mark the moonlanding. A crowd of several thousand assembled in Central Park for a city-sponsored event dubbed the "Moon-In," featuring three oversized television screens (3 × 4 meters) and accompanied by searchlights, balloons, dance performers in fantastic costume, clips from Buck Rogers movies, and musical acts including the proto-electronic band Silver Apples. The *Village Voice* called it "the world's largest outdoor multi-media program."[10] Because of heavy rain showers, the grounds had turned to mud, and barefooted hippies began to slide and dance wildly, strangely anticipating the Woodstock festival that would take place only a few weeks later. After many hours, as the hatch of the Landing Module finally opened and Neil Armstrong descended, the mammoth screens displayed the unprecedented caption: "Live from the surface of the moon." As has been noted, this collective witnessing was itself a marvel of modern technology: "The achievement of broadcasting live television from the Moon was nearly as astonishing as landing there."[11]

The year before, the Italian artist Fabio Mauri had staged a lunar voyage of his own with his installation *Luna*, a walk-in environment that placed the spectator in a simulated lunar environment, ultimately raising questions about the cultural, political, and artistic meanings of space exploration as vicarious media spectacle. Held at the Galleria della Tartaruga in Rome, Mauri's ephemeral work consisted of a space within the gallery, which could be entered through two lozenge-shaped openings that have been compared to spaceship hatches. Inside, against black walls, a foot-deep layer of white Styrofoam pellets covered the floor of the otherwise empty chamber, lit obliquely through the doors so as to create a cratered lunar surface. Spectators assume the role of astronauts walking on an unfamiliar and uncanny substrate. Wrote one critic, "It almost seems as if one loses one's gravitational weight."[12] The experience was at once psychological and, as critics have pointed out, sociological. Carolyn Christov-Bakargiev has suggested that Mauri's black-and-white space was a critical intervention in the much-heralded moonlanding as a televised media event, "as if it were possible to make a real experience out of a mediated experience, physically and literally to enter into a monitor."[13]

Perhaps no one understood the mediated nature of space exploration more than Robert Rauschenberg in his 1969 print series *Stoned Moon* (pp. 94-95). Rauschenberg, a space enthusiast (in 1965 he named his dog Laika in honor of the first animal ever to orbit the earth), was invited to witness the Apollo 11 launch at Cape Kennedy in Florida. He arranged with Kenneth Tyler at Gemini G.E.L. in Los Angeles to produce a suite of lithographs in response. Rauschenberg secured access to an array of images from NASA's photographic archive – blueprints, publicity photographs, maps, and various illustrations. He meanwhile culled additional pictures relating to human flight (the Wright brothers, Charles Lindberg) and to the natural environment and local culture in Florida (palm trees, birds, oranges).

The title of the series, *Stoned Moon*, refers at once to the lunar surface, to the lithographic stones used to produce these prints, and to the overall sense of intoxication induced by the lunar program as Rauschenberg represented it, its overwhelming nature. Layering pictures of astronauts, rockets, technical diagrams, maps, and images from nature along with abstract marks in crayon and tusche, the prints, some monumental in scale, are visually complex. Rauschenberg's free-floating and sometimes indistinct imagery can be hard to interpret thematically as well as to decipher visually. As Jaklyn Babington has observed, "The 'static' that we see in the artist's series of prints mimics the reception of live recordings from the moon to so many small black-and-white television screens and radio sets around the globe."[14]

The human figure is central to Rauschnberg's prints, even as he resists the familiar heroic image of the astronaut's face and figure. Instead of romanticizing the astronaut, Rauschenberg locates the figure in a fragmented environment. In an unpublished analysis, Michael Crichton identifies the complexity, ambivalence, and ambiguity with which Rauschenberg represents the Apollo program – a departure, as Crichton sees it, from the more triumphalist and naturalistic approach of other NASA-affiliated artists at that time. An example of this ambivalence and ambiguity is *White Walk*, which takes a famous image of American astronaut Ed White's pioneering 1965 space walk and repeats it in multiple, overlapping orientations while collaging additional images of control panels and technical apparatus along with the artist's marks. Ultimately, for Crichton, Rauschenberg was highlighting the "technical complexity of this romantic undertaking," the inescapable irony that "Man needs all these machines to float in space."[15]

A number of the *Stoned Moon* prints contain the image of Buzz Aldrin's footprint on the moon, evidence of the human body's contact with the powdery lunar surface, and one of the most widely reproduced pictures to come out of what was a highly productive image-making enterprise (space exploration has always been the occasion for generating pictures). The Apollo 11 mission was extensively photographed, filmed, and, perhaps most important, televised. At a time when mass media were on the rise world-wide (and when artists were freely blurring boundaries between art forms old and new), the moonlanding was a global multimedia event, combining aspects of mass spectacle and audience participation. It represented not only a new reality but also a new, space-age sensibility, widely reflected in trends in design and architecture and in contemporary entertainment such as Kubrick's *2001* and David Bowie's *Space Oddity*. If the limits of technology were expanding rapidly, so, in the endless and free-floating realm of outer space, were the limits of human consciousness.

1 Michael Kirby, *Happenings: An Illustrated Anthology*, New York: E.P. Dutton and Co., 1965, 137-38.
2 David G. Hartwell, "Introduction," in Robert Heinlein, *Destination Moon*, Boston: Gregg Press, 1979, vii.
3 Raffaele Carrieri, "Fontana ha toccato la luna," *Tempo*, February 19-26, 1949, p. 28.
4 Richard Hamilton, "Urbane Image," *Living Arts*, no. 2, 1963, p. 49.
5 *Richard Hamilton*, London: Tate Gallery, 1992, p. 154.
6 Yves Klein, *Conférence de la Sorbonne*, 3 juin 1959, Paris: Galerie Montaigne, [1992], n.p.
7 Pierre Restany, *Yves Klein*, trans. John Shepley, New York: Harry N. Abrams, 1982, p. 227.
8 Guido Le Noci, "Nota della galleria," in *Yves Klein le Monochrome: Il Nuovo Realismo del Colore*, Milan: Galleria Apollinaire, 1961, n.p.
9 Ciara Moloney, *Fly Me to the Moon*, Oxford: Modern Art Oxford, 2015, p. 47.
10 Steve Lerner, "The Age of Lunacy on a Muddy Meadow," *Village Voice* XIV:41, July 24, 1969, p. 1.
11 David Meerman Scott and Richard Jurek, *Marketing the Moon: The Selling of the Apollo Lunar Program,* Cambridge, Massachusetts, and London: MIT Press, 2013, xi.
12 Achille Bonito Oliva, "Il teatro delle mostre," *Sipario*, July 1968, p. 9.
13 Carolyn Christov-Bakargiev, "Nello schermo: insonnia per diverse forme contrarie di universo," in *Fabio Mauri: Opere e Azioni, 1954–1994*, Rome: Galleria Nazionale d'Arte Moderna e Contemporanea; Mondadori, 1994, p. 78.
14 Jaklyn Babington, *Robert Rauschenberg: Stoned Moon,* Canberra: National Gallery of Australia, p. 9.
15 Michael Crichton, draft introduction to *Stoned Moon Book* (unpublished), Kenneth Tyler Collection, National Gallery of Australia, Canberra, p. 21.

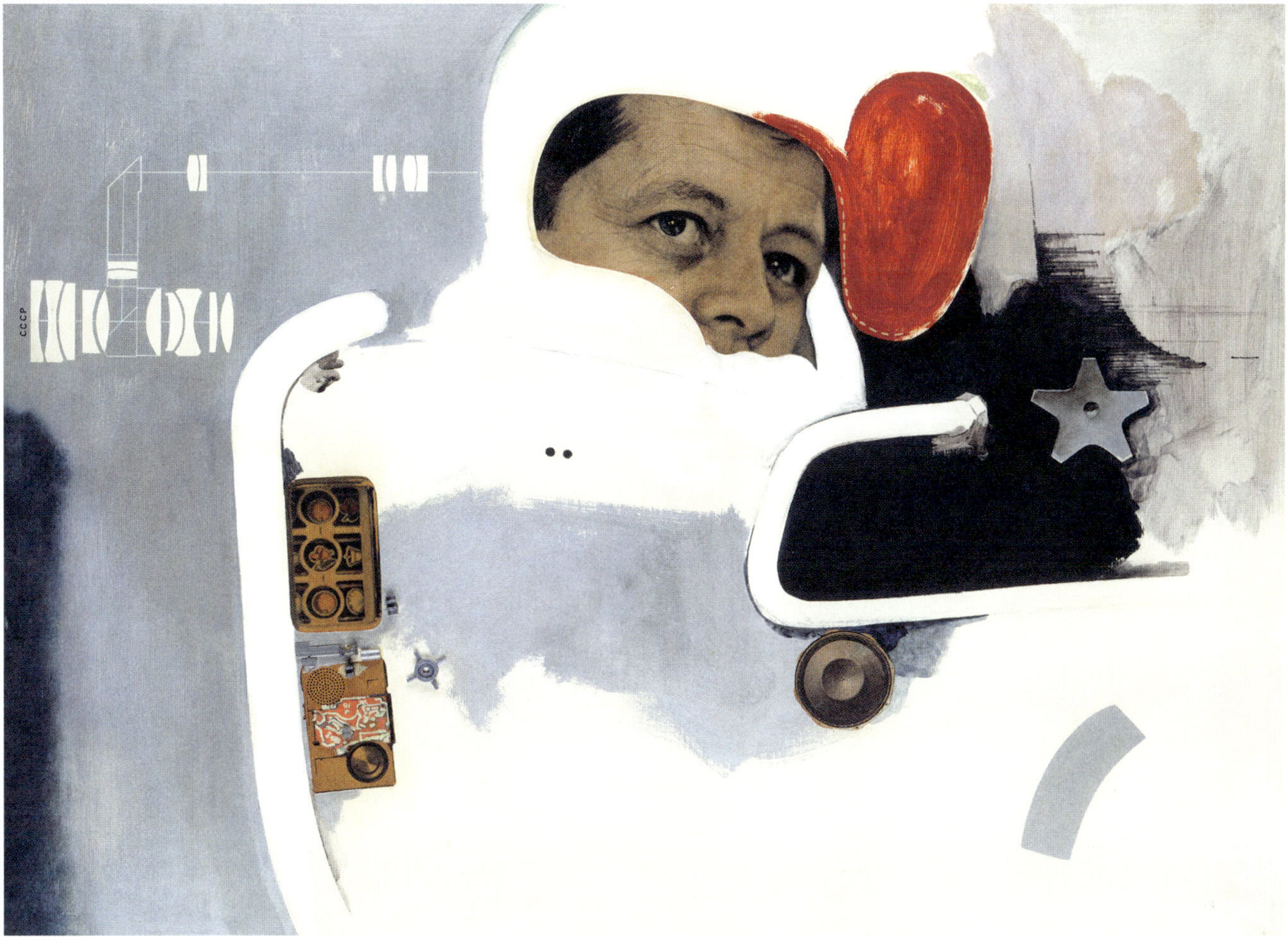

cat. 64
Richard Hamilton
Towards a definitive statement on the coming trends in menswear and accessories (a) Together let us explore the stars, 1962

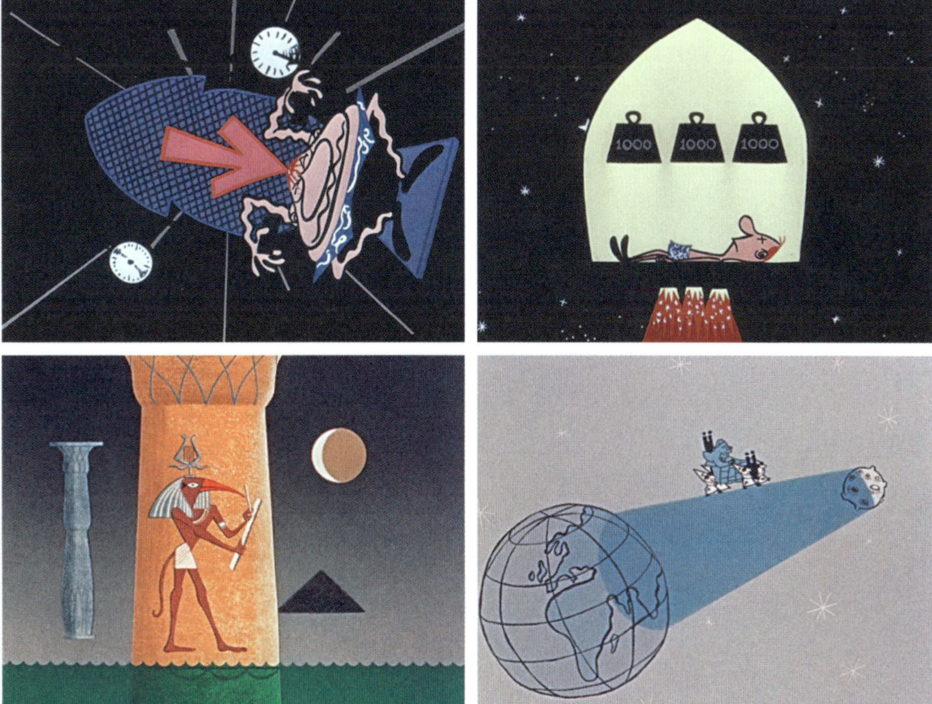

cat. 206-207
Stills from Walt Disney Studios' *Man and the Moon*, 1955
and *Man in Space*, 1955

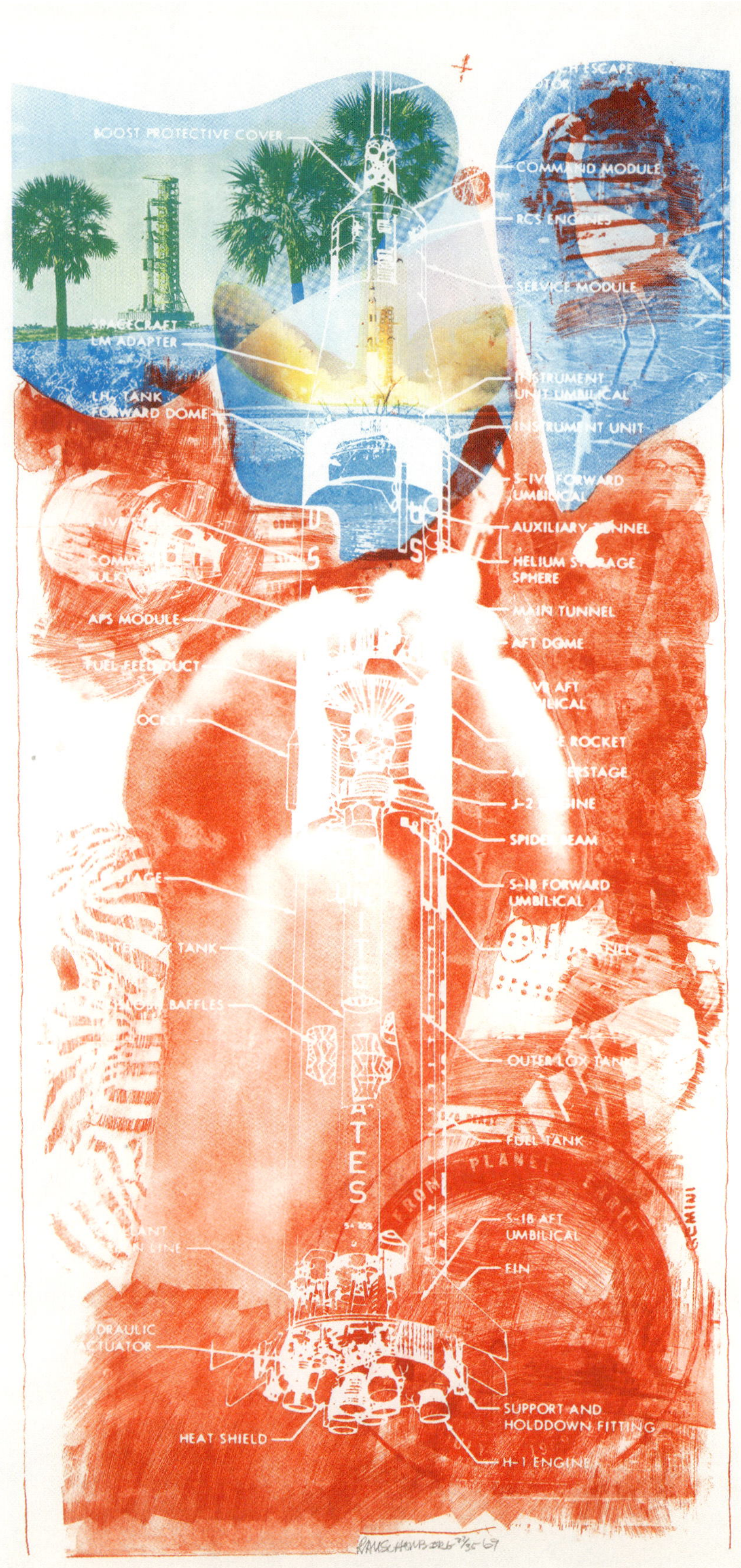

cat. 126
Robert Rauschenberg
Cover of *Stoned Moon Book*, 1970

cat. 133
Robert Rauschenberg
Page from *Stoned Moon Book*, 1970

cat. 113
Robert Rauschenberg
Sky Garden, 1969

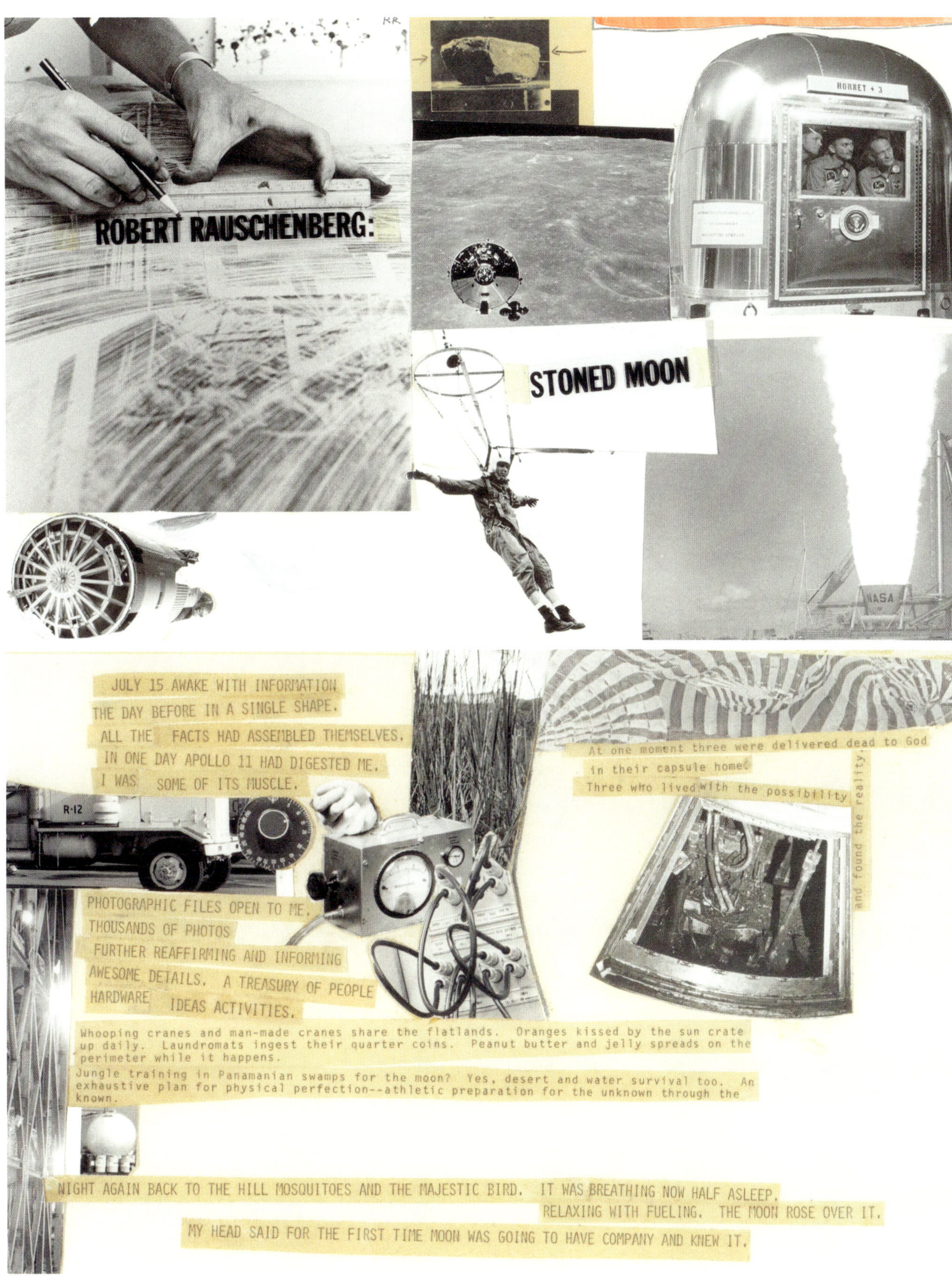
KR
ROBERT RAUSCHENBERG:
HORNET + 3
STONED MOON
NASA
JULY 15 AWAKE WITH INFORMATION
THE DAY BEFORE IN A SINGLE SHAPE.
ALL THE FACTS HAD ASSEMBLED THEMSELVES.
IN ONE DAY APOLLO 11 HAD DIGESTED ME.
I WAS SOME OF ITS MUSCLE.
At one moment three were delivered dead to God
in their capsule home.
Three who lived with the possibility
and found the reality.
R-12
PHOTOGRAPHIC FILES OPEN TO ME.
THOUSANDS OF PHOTOS
FURTHER REAFFIRMING AND INFORMING
AWESOME DETAILS. A TREASURY OF PEOPLE
HARDWARE IDEAS ACTIVITIES.
Whooping cranes and man-made cranes share the flatlands. Oranges kissed by the sun crate up daily. Laundromats ingest their quarter coins. Peanut butter and jelly spreads on the perimeter while it happens.
Jungle training in Panamanian swamps for the moon? Yes, desert and water survival too. An exhaustive plan for physical perfection--athletic preparation for the unknown through the known.
NIGHT AGAIN BACK TO THE HILL MOSQUITOES AND THE MAJESTIC BIRD. IT WAS BREATHING NOW HALF ASLEEP.
RELAXING WITH FUELING. THE MOON ROSE OVER IT.
MY HEAD SAID FOR THE FIRST TIME MOON WAS GOING TO HAVE COMPANY AND KNEW IT.

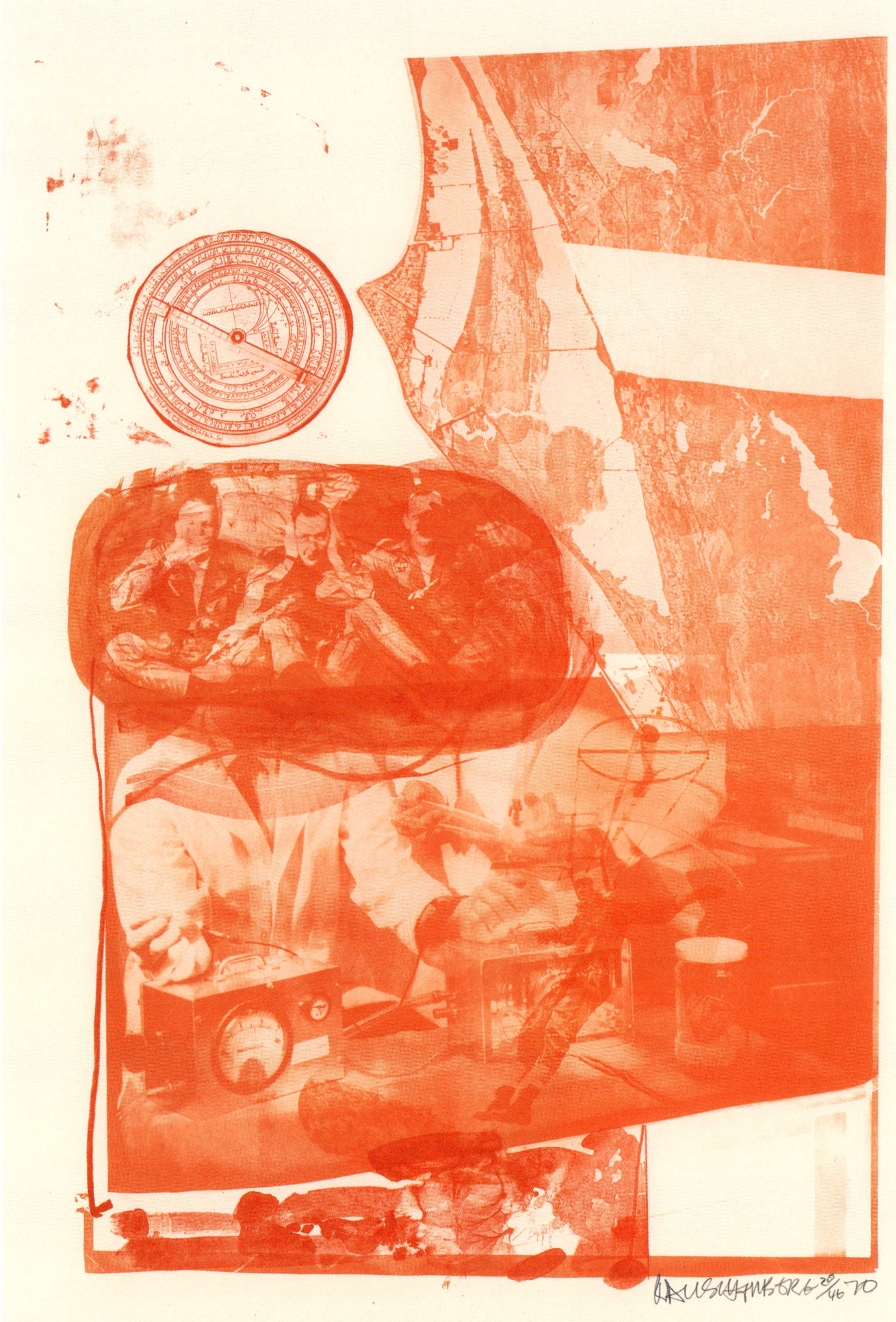

cat. 124
Robert Rauschenberg
Hybrid (Stoned Moon), 1970

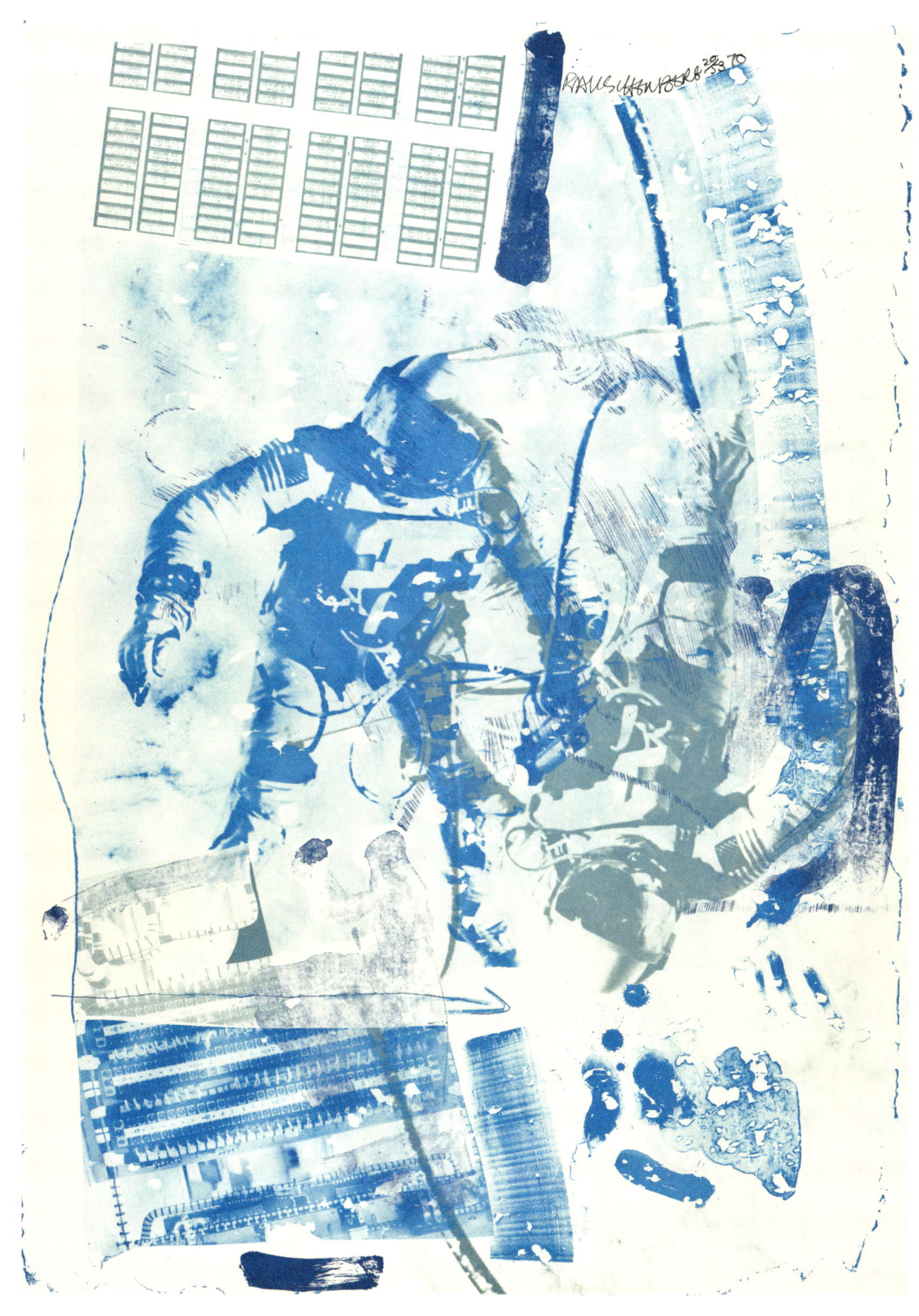

cat. 138
Robert Rauschenberg
White Walk (Stoned Moon), 1970

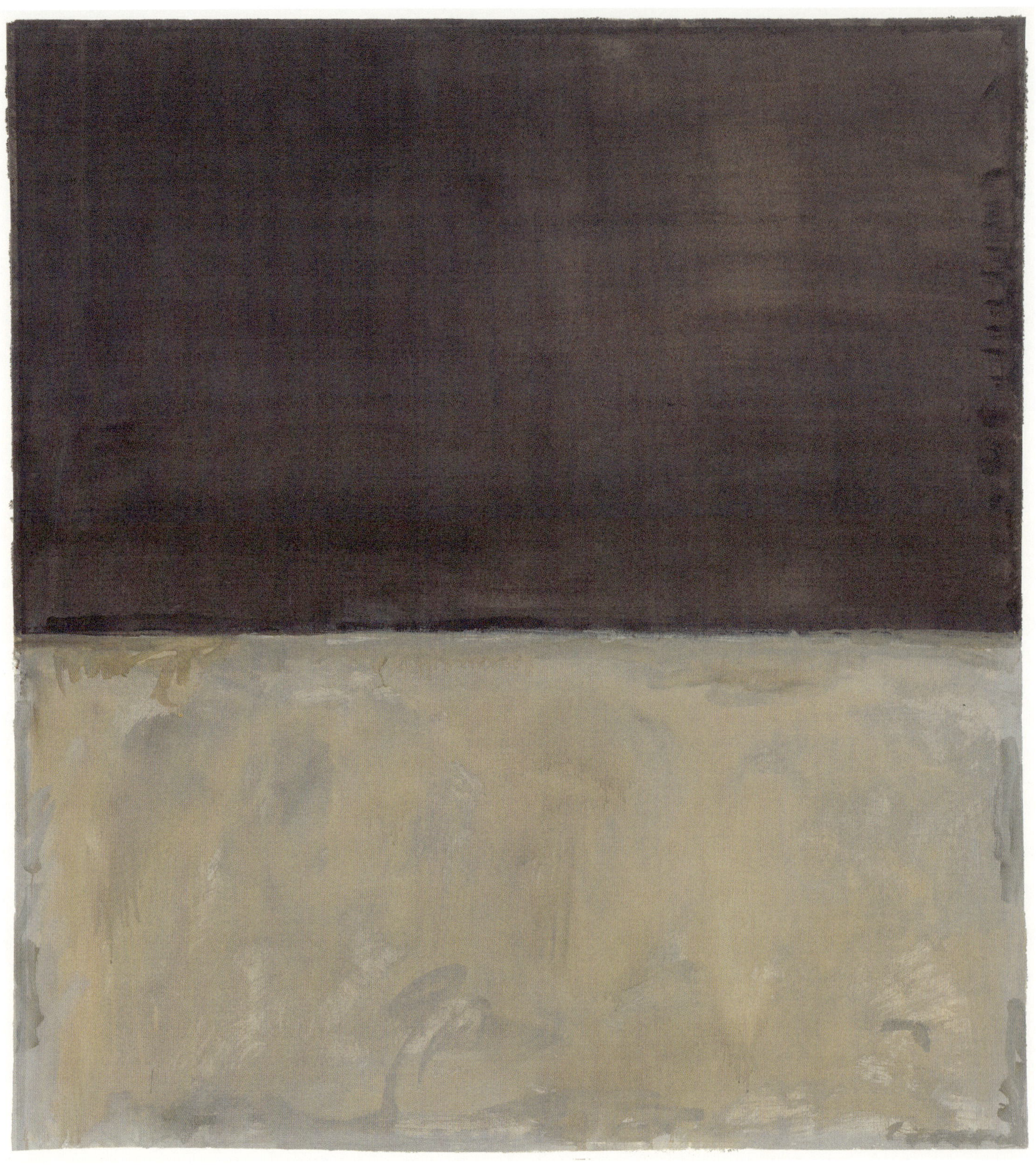

cat. 142
Mark Rothko
Untitled, 1969
Uden titel

cat. 158
Alan Shepard
From the Mauro landing site against brilliant Sun glare, Apollo 14, February 1971

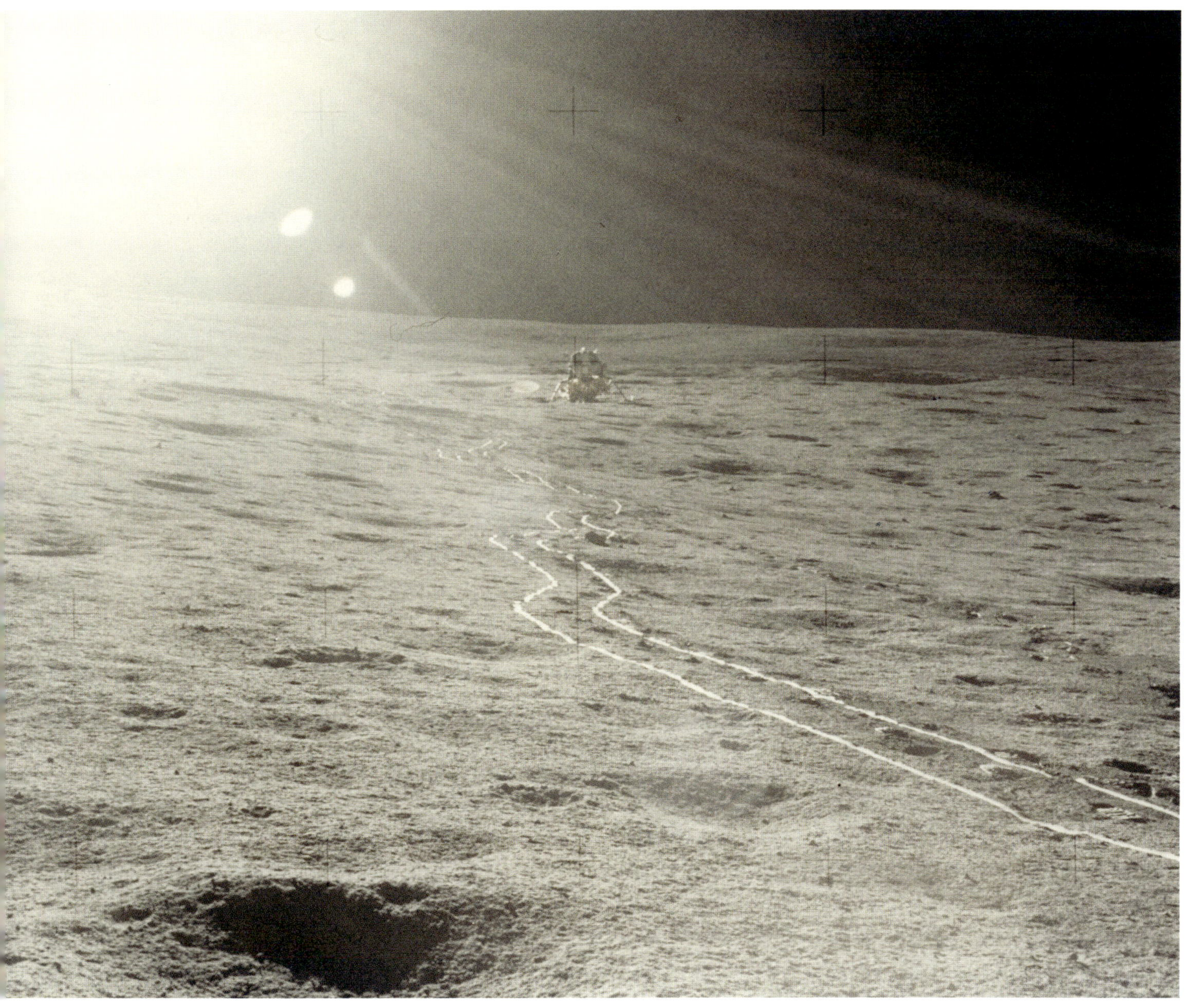

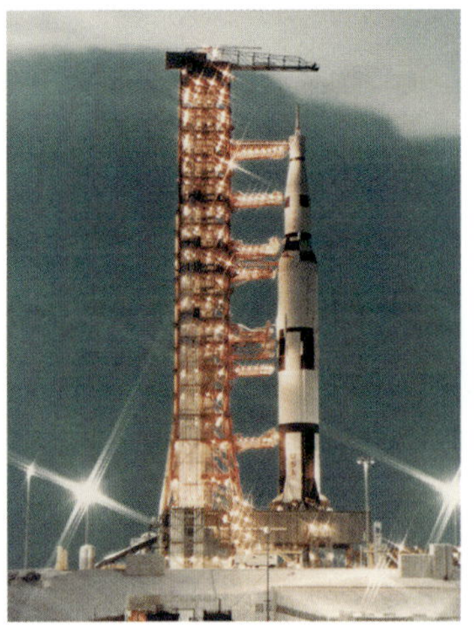

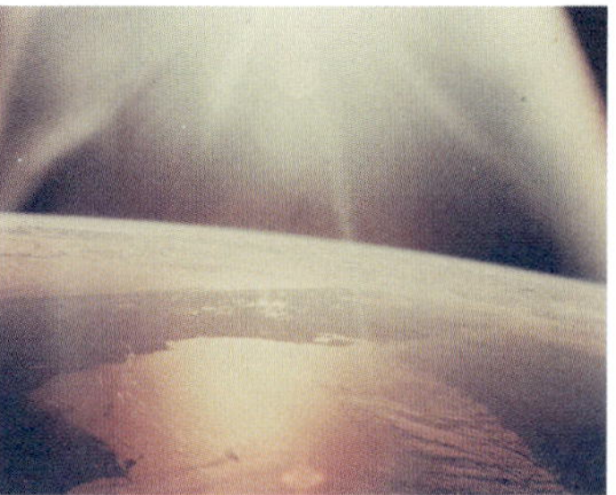

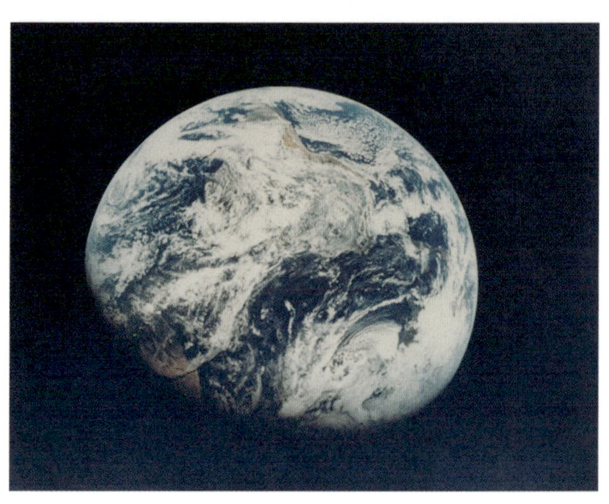

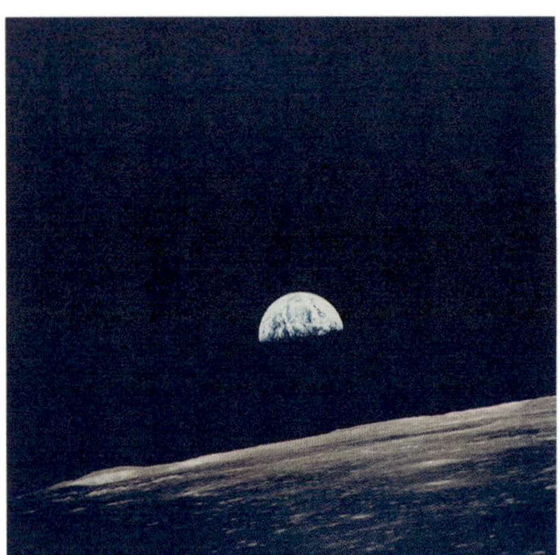

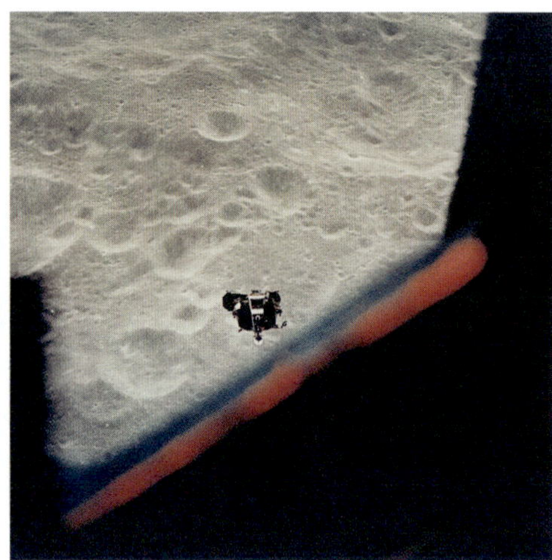

Above:
cat. 182
Apollo 4, 1966
cat. 40
Apollo 7, 1968
cat. 14
Apollo 8, 1968

Left:
cat. 188
Apollo 12, 1969
cat. 215
Apollo 10, 1969

Below:
cat. 6
Apollo 11, 1969
cat. 185
Apollo 11, 1969
cat. 18
Apollo 11, 1969

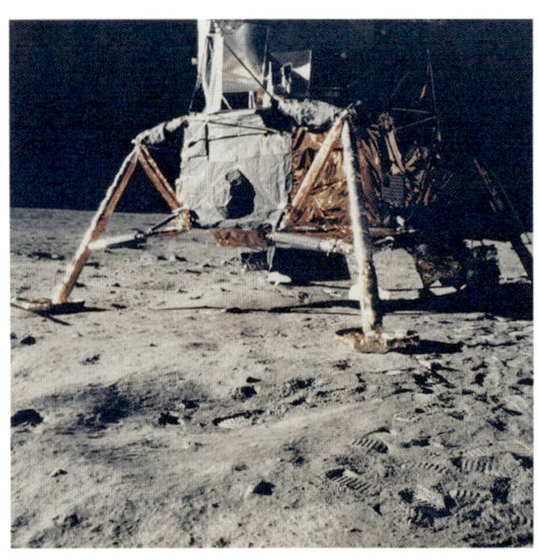

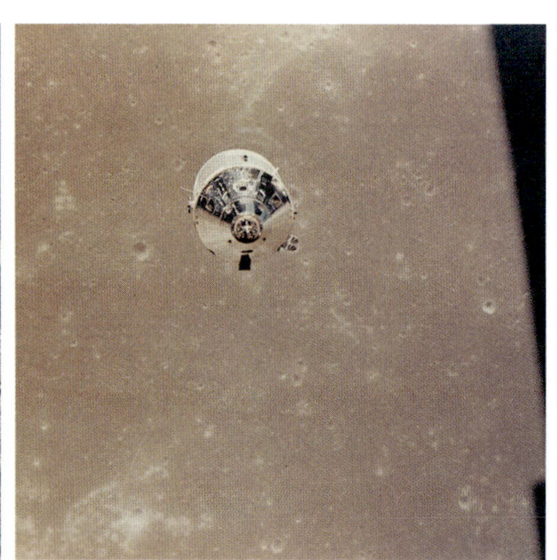

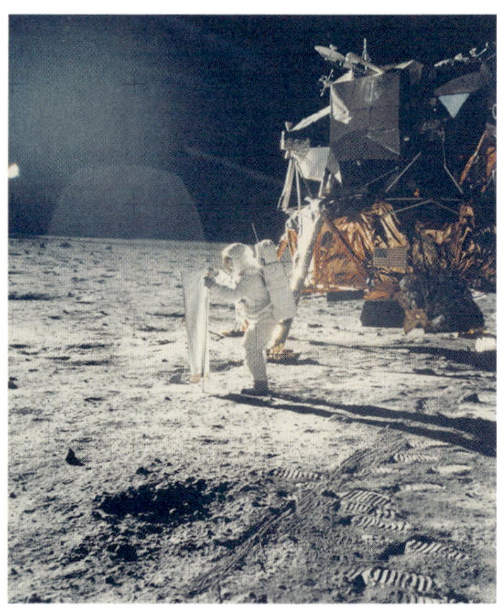

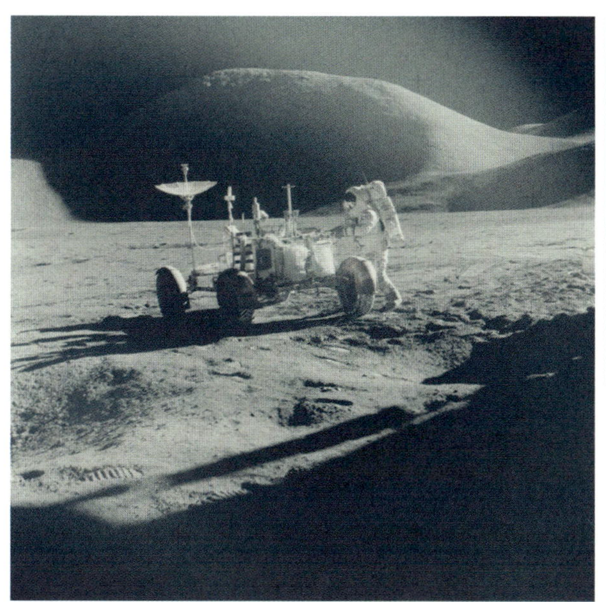

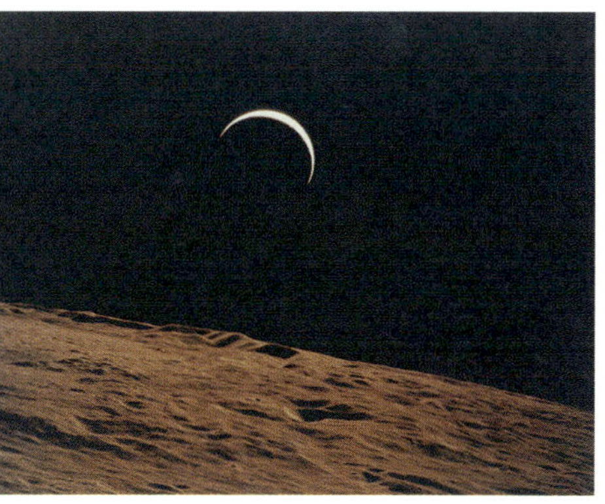

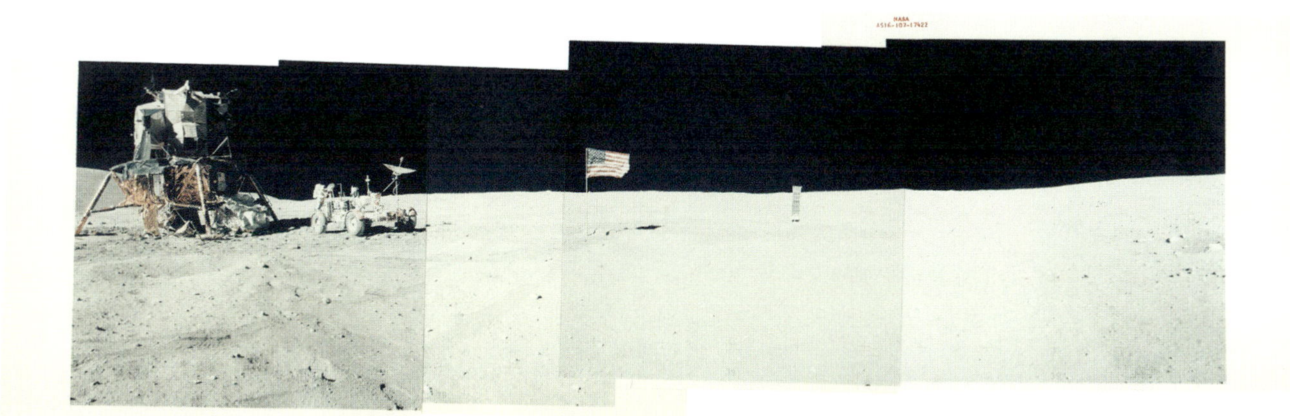

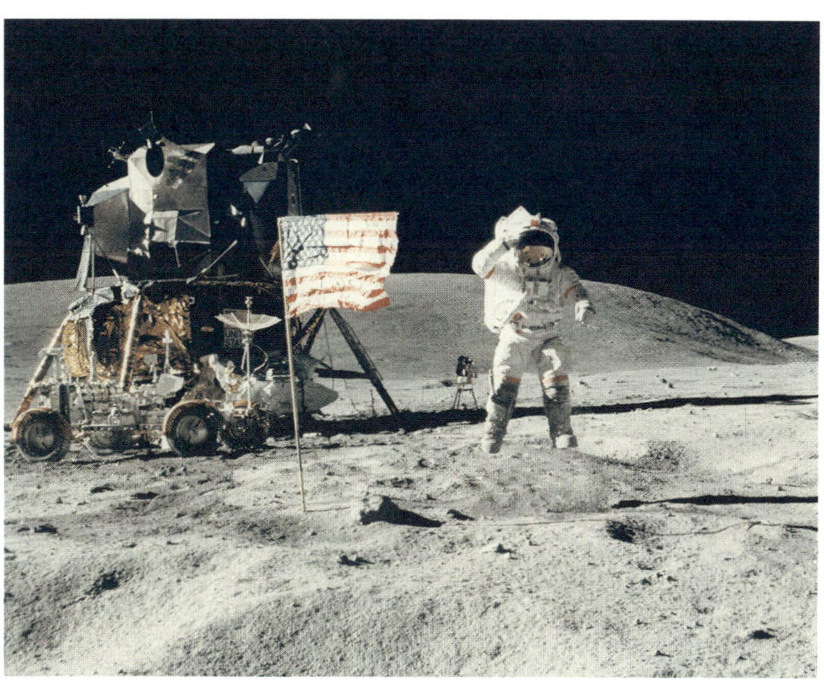

Above:
cat. 154
Apollo 15, 1971
cat. 210
Apollo 15, 1971
cat. 50
Apollo 16, 1972

Right:
cat. 49
Apollo 16, 1972

Right:
cat. 87
Gemini 4, 1965
cat. 89
Gemini 4, 1965
cat. 180
Gemini 4, 1965
cat. 88
Gemini 4, 1965

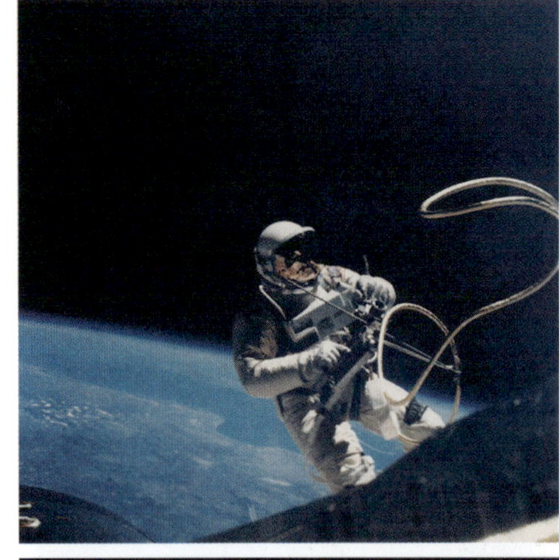

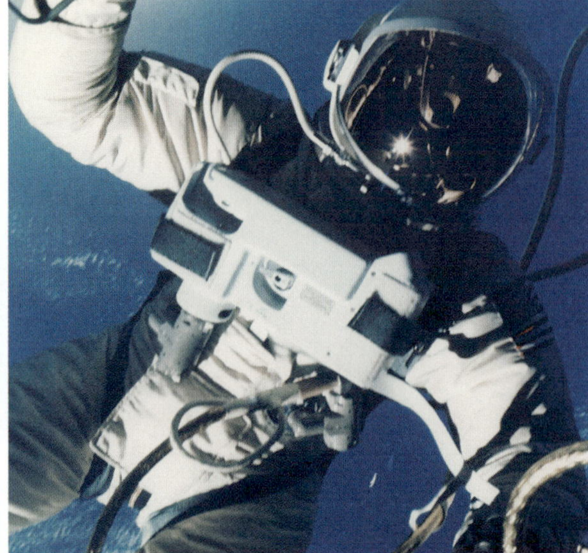

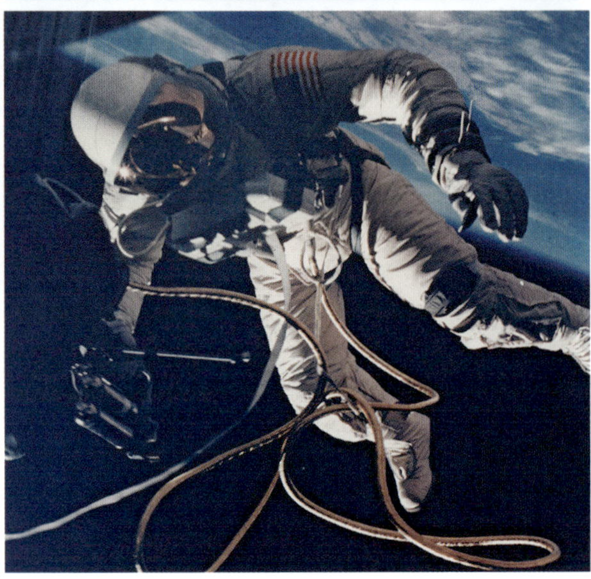

Below:
cat. 181
Gemini 5, 1965
cat. 30
Gemini 11, 1966

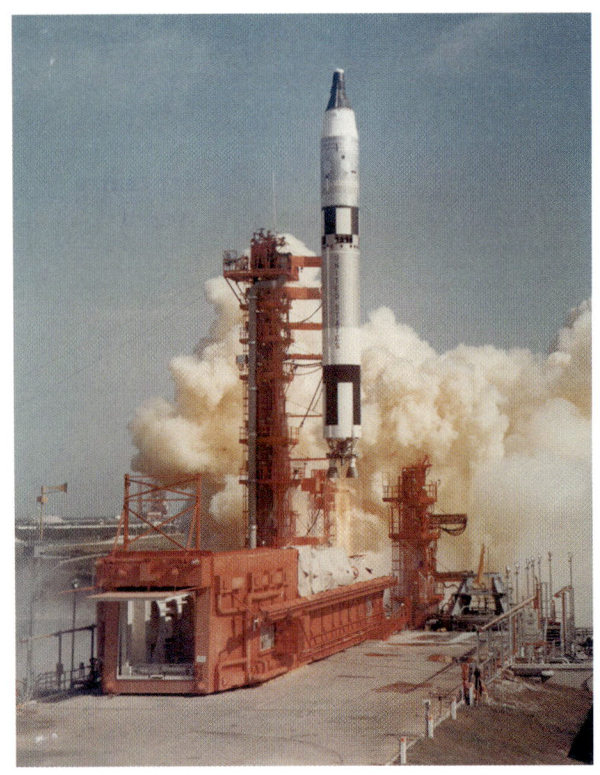

Right:
cat. 68
Apollo 15, 1971
cat. 156
Apollo 15, 1971
cat. 211
Apollo 15, 1971

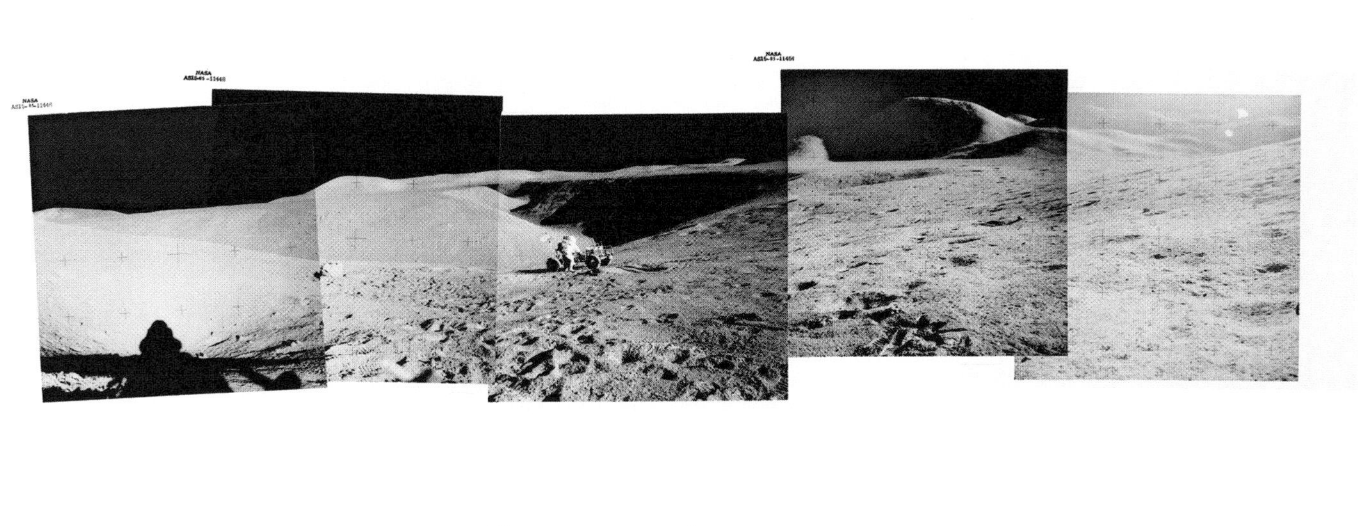

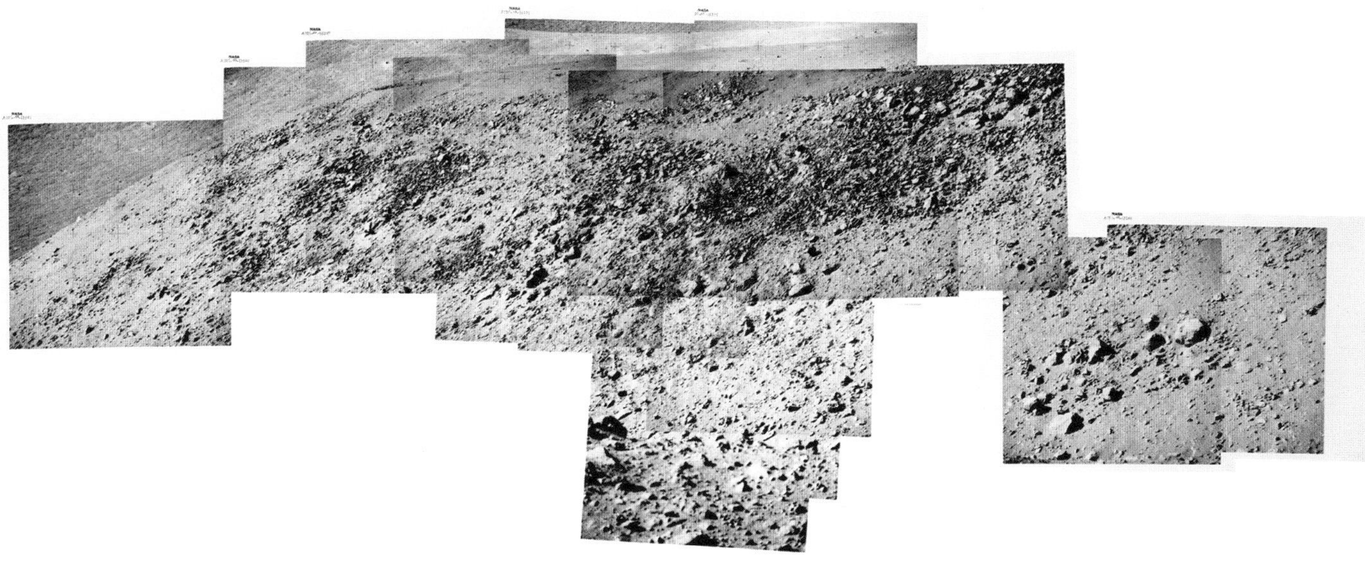

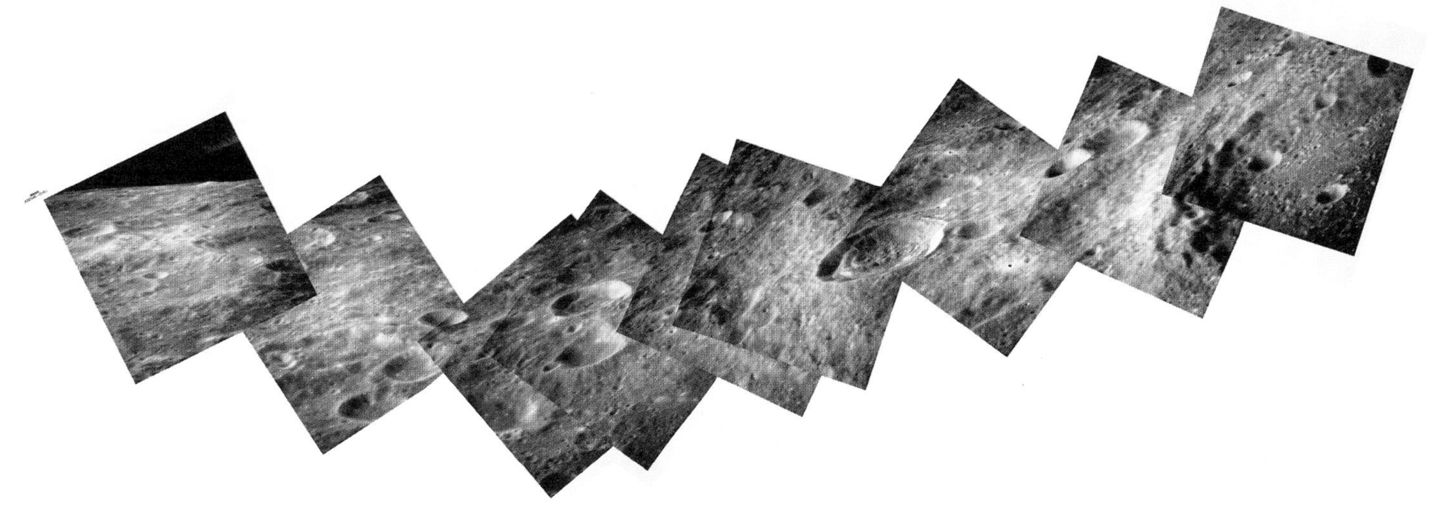

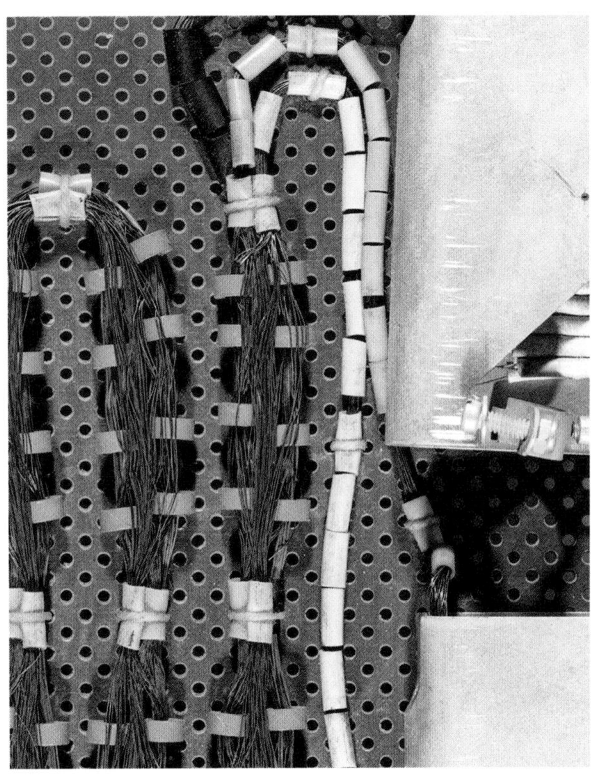

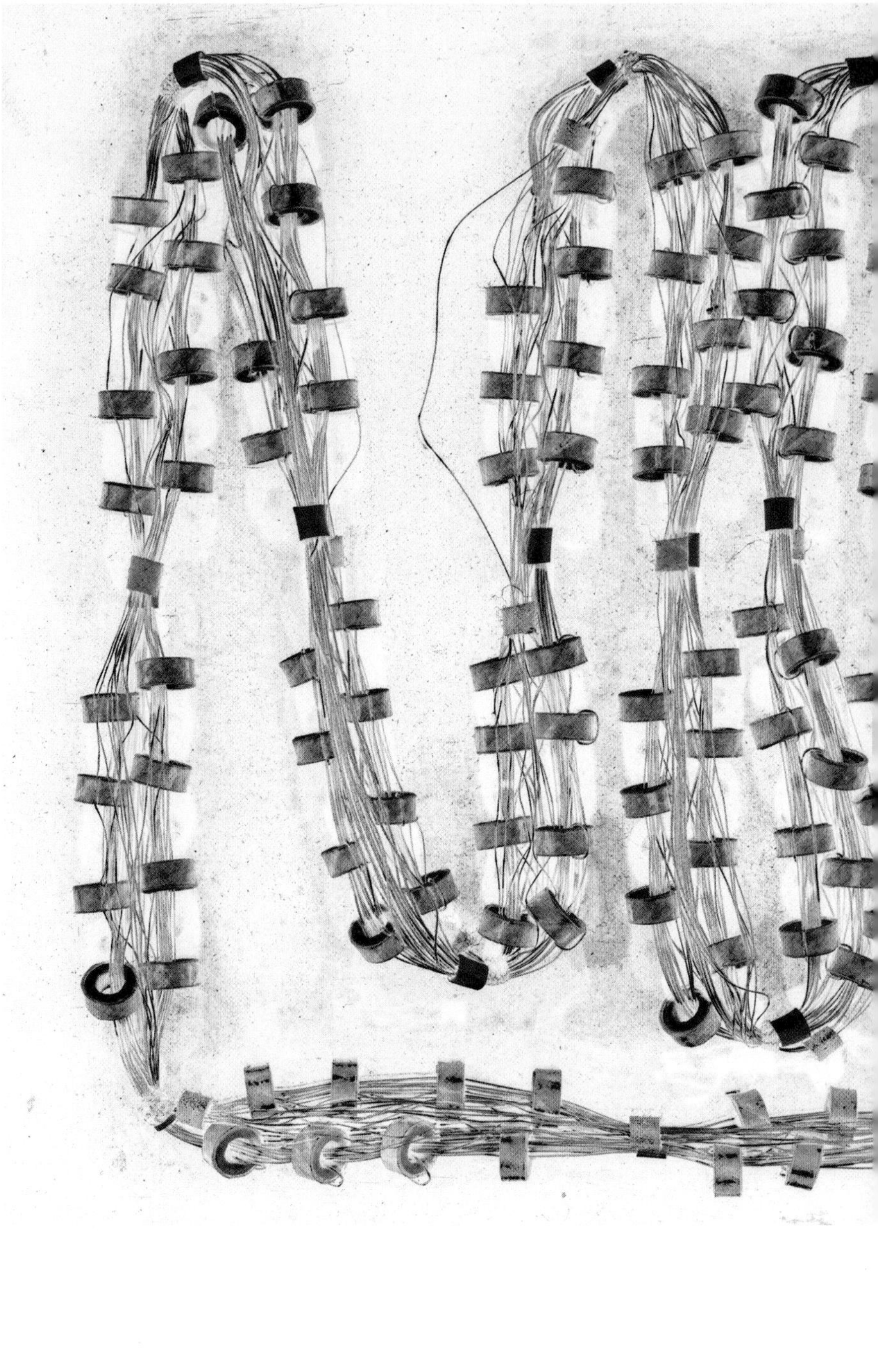

cat. 29
Nanna Debois Buhl
Moon Memory, 2018

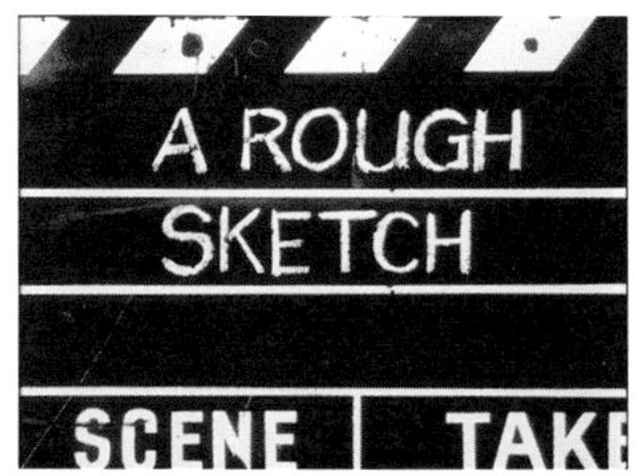

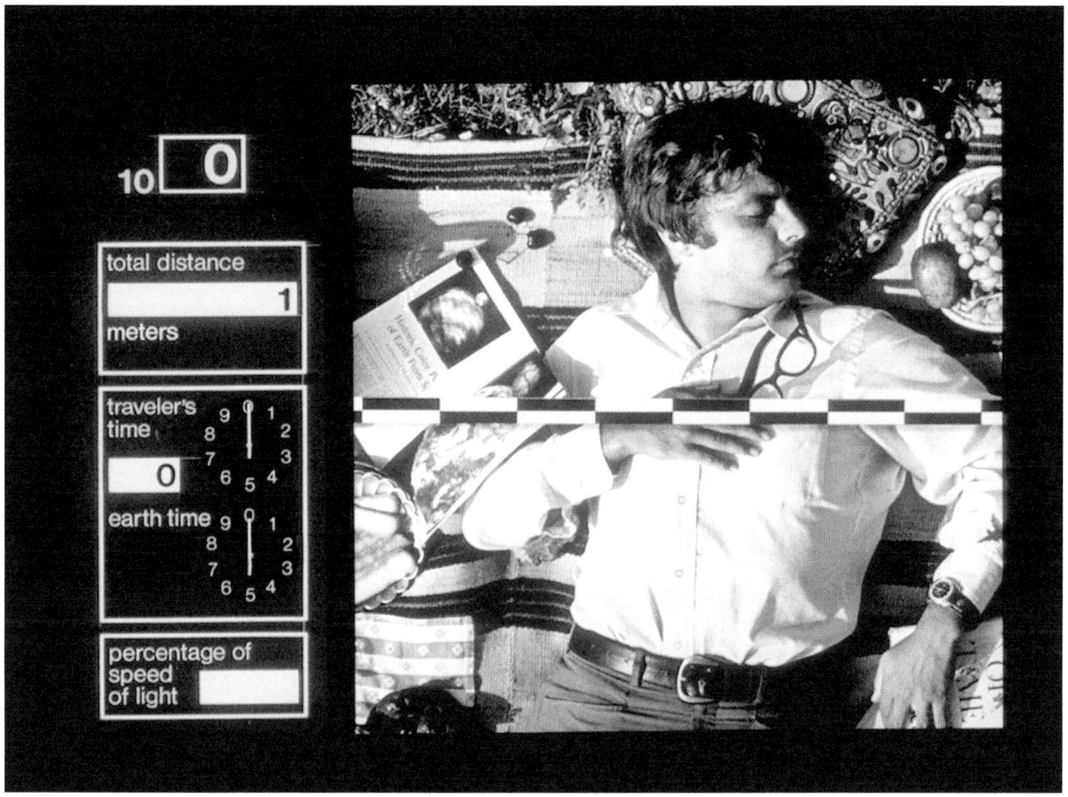

cat. 52
Charles Eames
A Rough Sketch for a Proposed Film Dealing with the Powers of Ten and the Relative Size of Things in the Universe, 1968

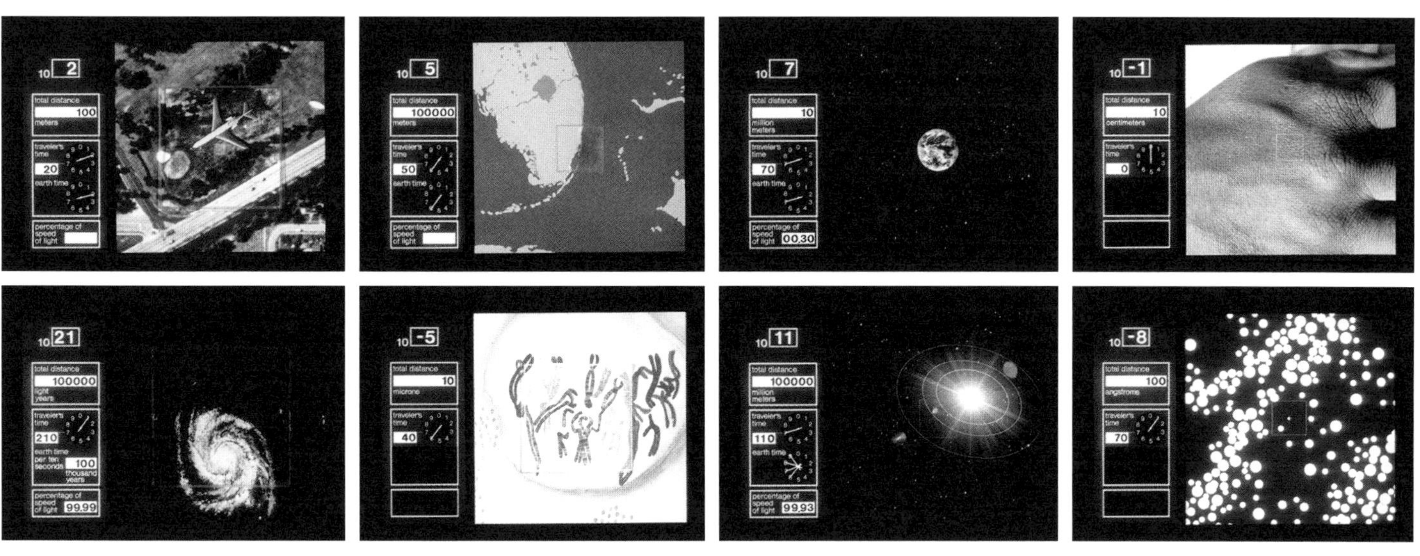

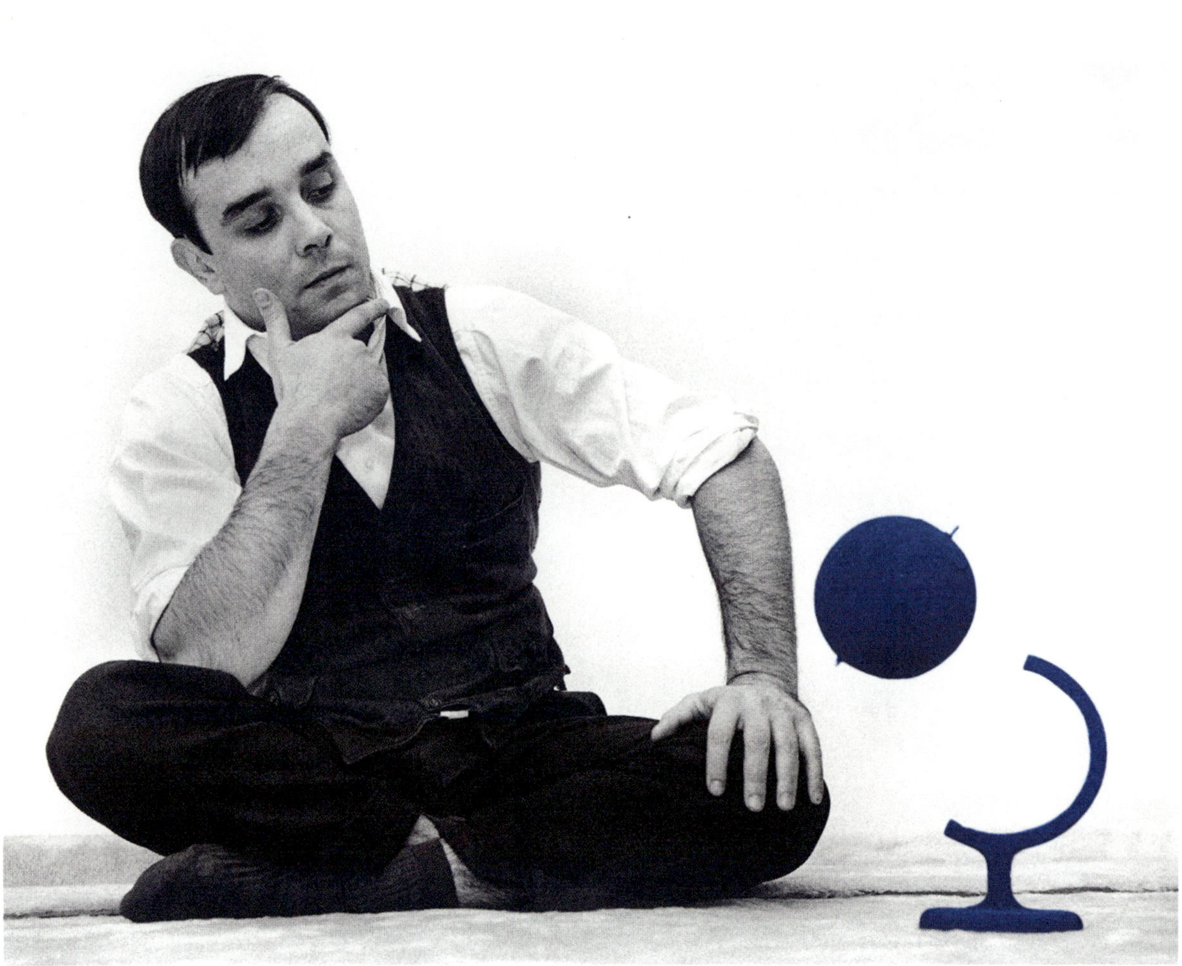

cat. 159
Harry Shunk & Janos Kender
Yves Klein with the Blue Globe (RP 7), 1961

Right:
cat. 74
Yves Klein
Untitled White Planetary Relief (RP 12), c. 1961

Next spread:

cat. 72
Yves Klein
Pink Planetary Relief "Lune II" (RP 21), 1961

cat. 73
Yves Klein
Planetary Relief "Region de Grenoble" (RP 10), 1961
(Posthumous edition c. 1990)

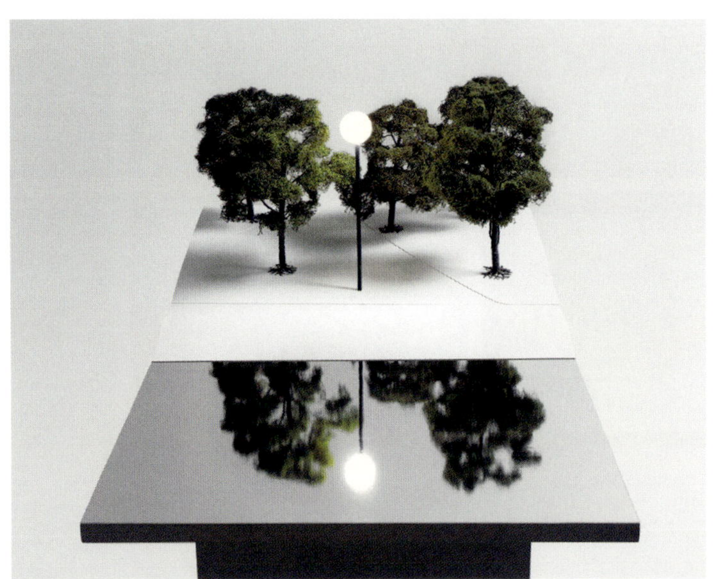

cat. 63
Isa Genzken
Vollmond, 2015 (Model)
Full Moon

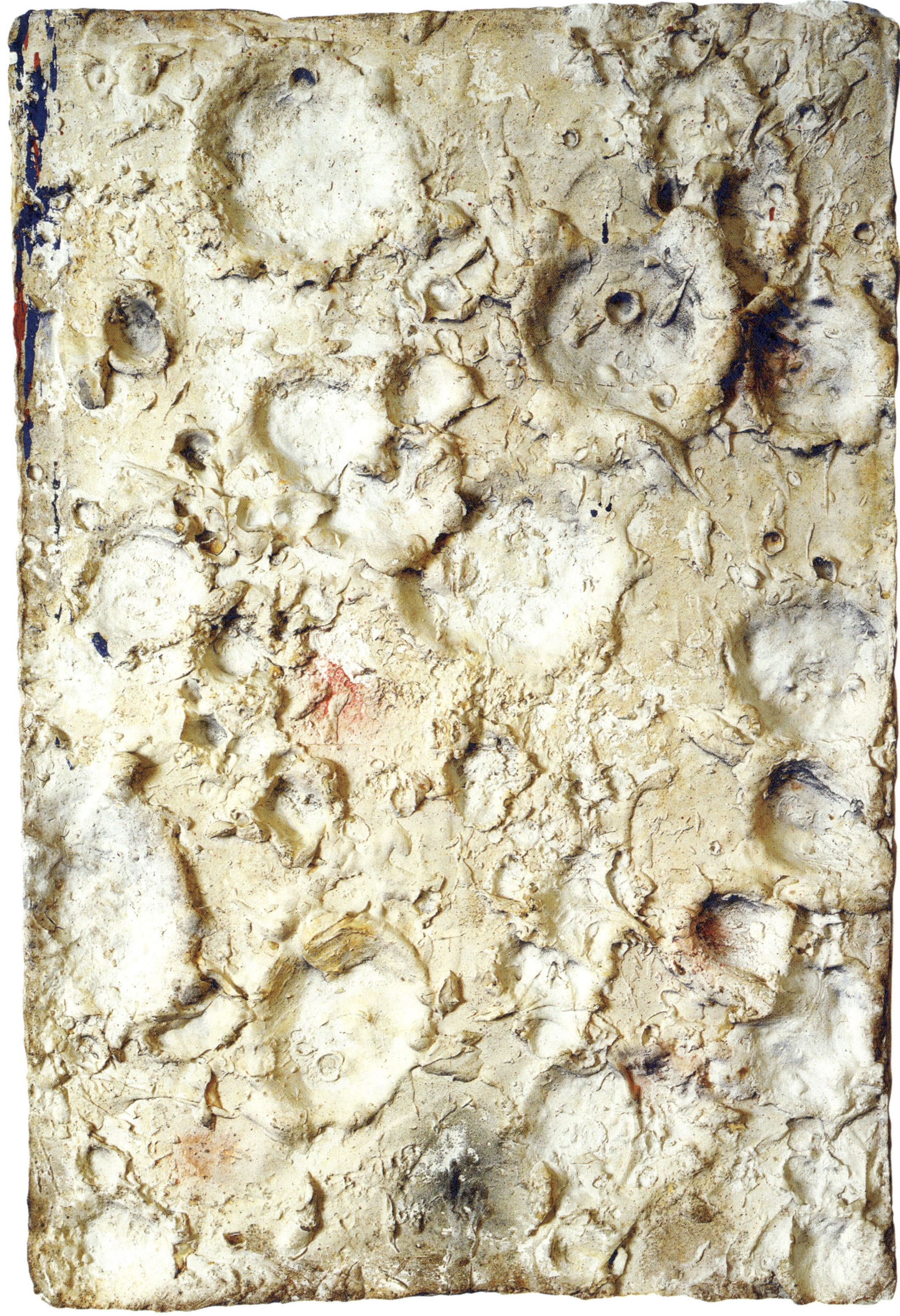

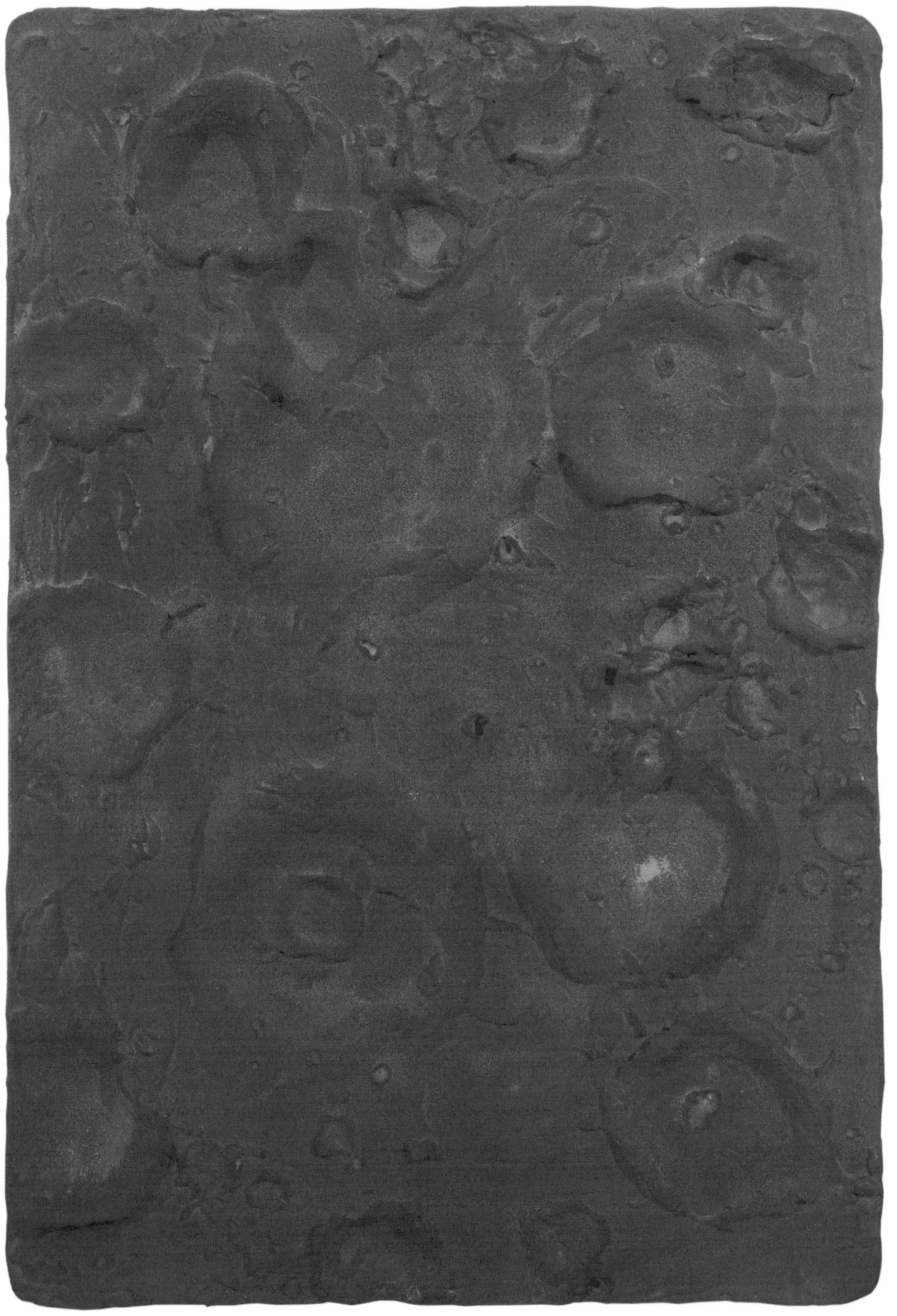

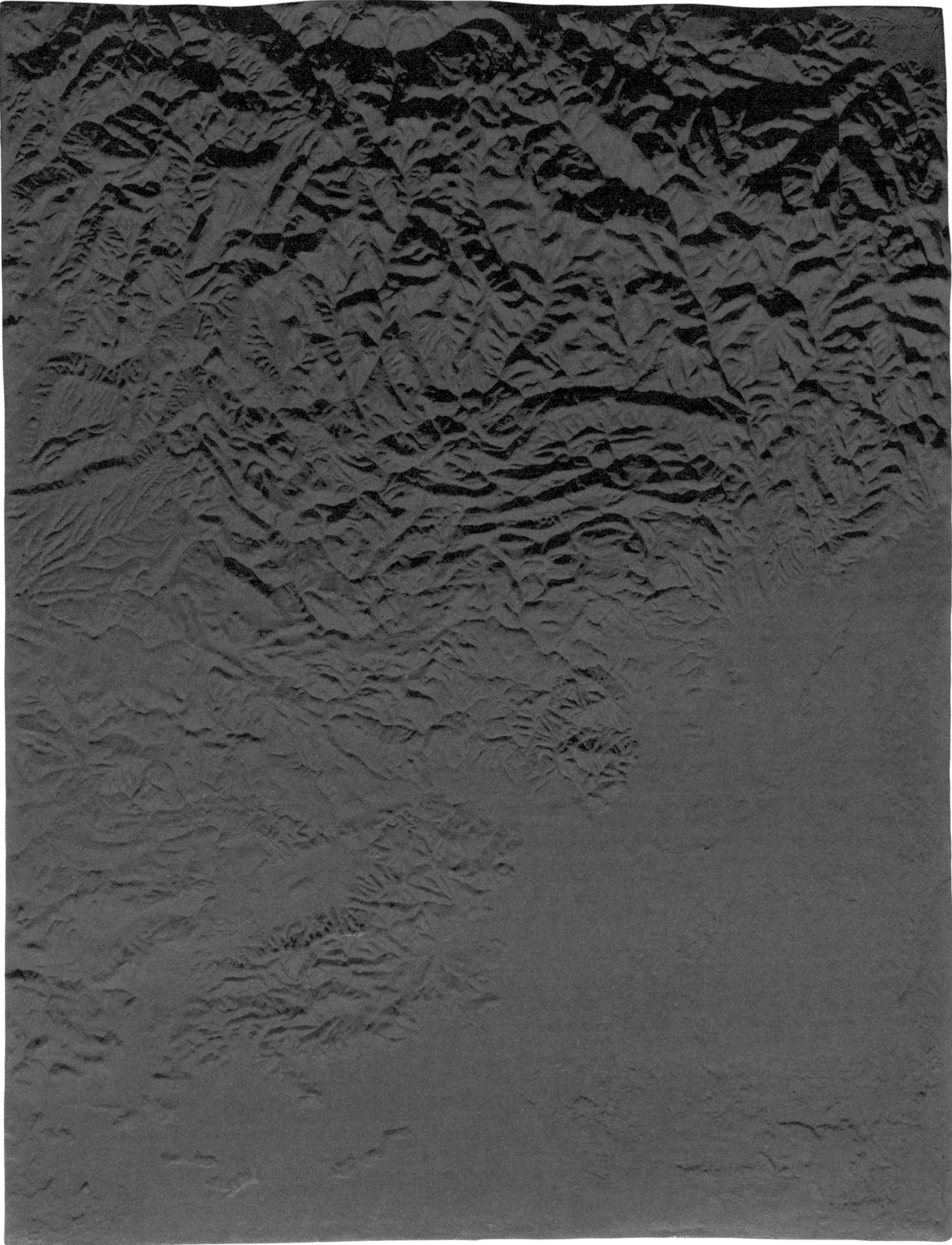

The Scientific Moon

by Anja C. Andersen

Astrophysicist Anja C. Andersen is Professor at the Niels Bohr Institute, University of Copenhagen. Her research is centred around cosmic dust and its important influence in relation to the formation of the stars and planets.

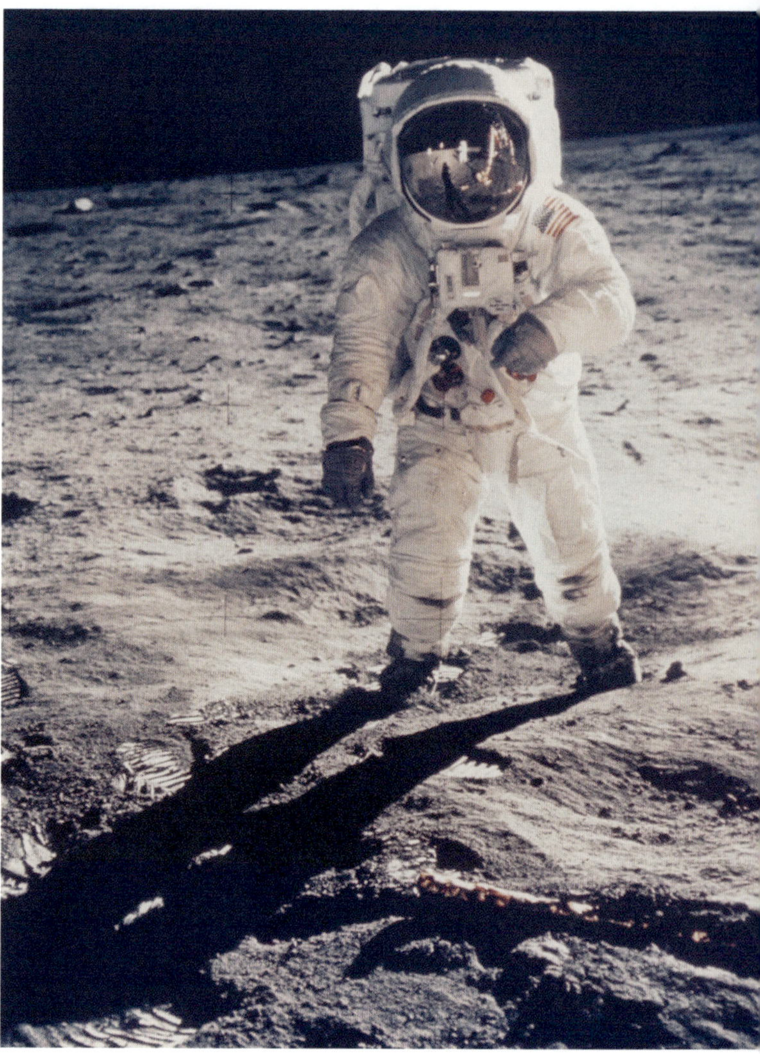

cat. 17
Neil Armstrong
Buzz Aldrin's gold-plated sun visor reflects the photographer and the Lunar Module, Apollo 11, July 1969

How I would love to walk on the moon. Experience for myself a weaker gravitational field than here on Earth. Jump lightly and effortlessly around in the lunar landscape, leaving deep footprints in the fine sand of the moon. Except maybe it would not be that effortless. I would have to wear a spacesuit, and they are far from easy to manoeuvre in. It is especially difficult to avoid kicking the inside of your legs as you walk. If I wanted to look behind me I would have to turn my entire body, because the helmet limits the field of view. And I would have to be careful, because a fall could be fatal. The spacesuit would make it difficult to get up, and even a tiny rip could be the end of me: air from the spacesuit would seep into the emptiness around me, and my oxygen reserve could fall to a life-threatening level before I made it back to the lander.

Crawling in and out of the vessel in a spacesuit is also incredibly difficult – at least as difficult as getting into the spacesuit, but not half as difficult as getting out of it again. The powdery moon dust sticks to the spacesuit when you are outside, but once you get inside the vessel again it gets everywhere. And that is dangerous, because not only do the mineral particles irritate the eyes and throat, they can also short-circuit the electricity supply.

When preparations were being made for the moon landings of the 1960s, however, spacesuits and moon dust were not the main concern. Back then nobody knew whether a spacecraft could even land on the moon, or whether it would be possible for it to *take off* again. Scientists at the time had no idea how hard the surface of the moon was, which is fascinating to think back on. Speculation centred on whether it was covered with a layer of dust – and the best estimates were that this layer was between two centimetres and two metres thick.

cat. 183
The first view of the Earth from the Moon, Frame 101, High Resolution, Lunar Orbiter I, 23 August 1966

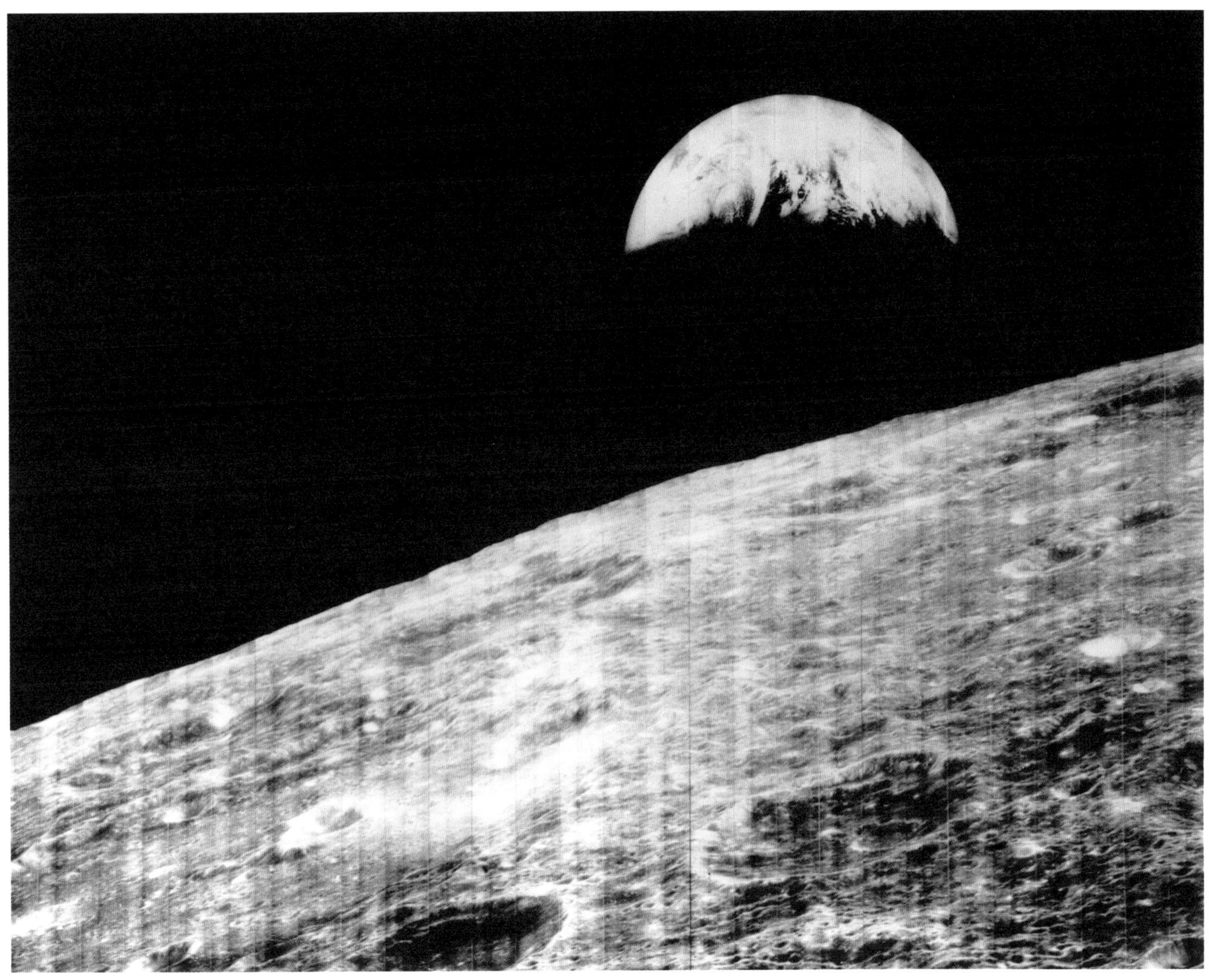

cat. 16
Neil Armstrong
Buzz Aldrin about to take his first step on the Moon, Apollo 11, July 1969

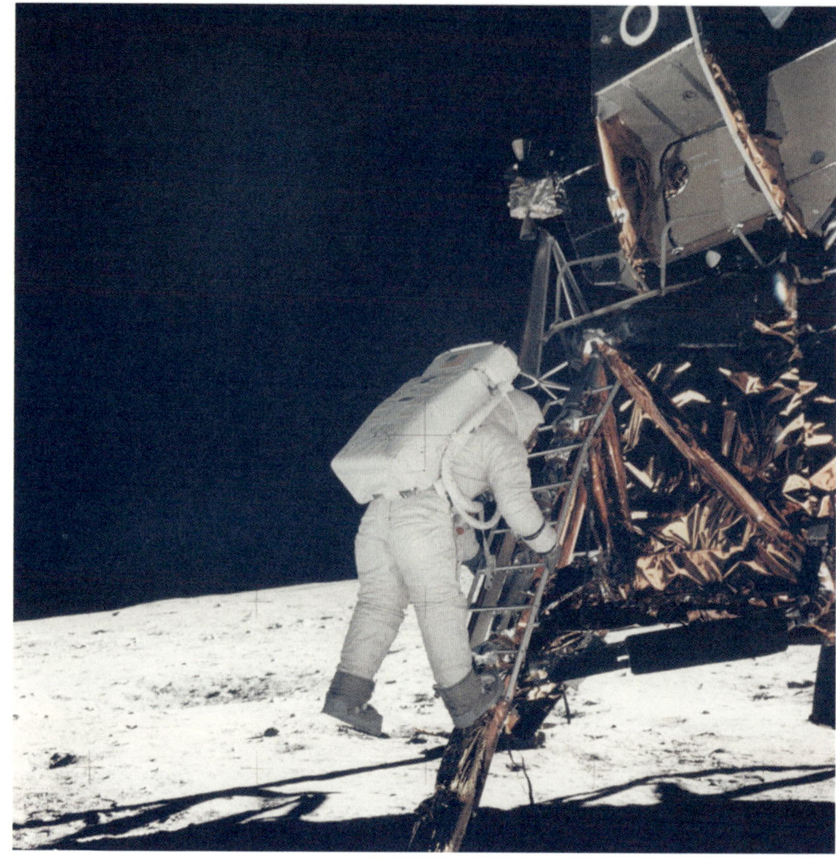

cat. 198
Ketophyllum (coral) from Gotland, which shows that the length of the year was 421 days around 420 million years ago

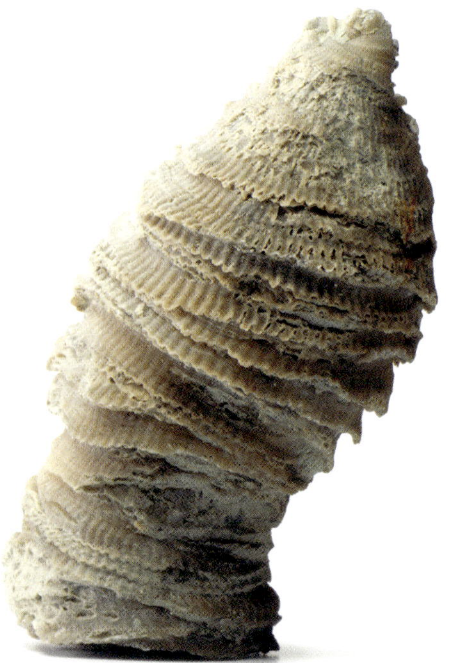

cat. 200
Lunar meteorite sample

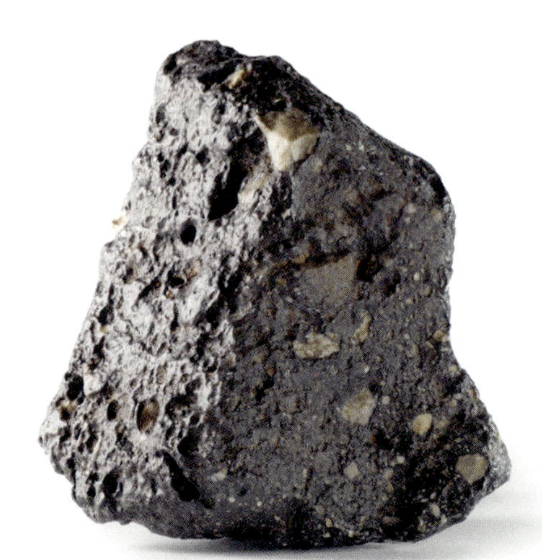

The Soviet Union was the first country – in 1966 – to make a successful landing on the moon with the Luna 9 space mission. But it was not exactly a smooth landing. The spacecraft reached the moon at a speed of 22 km/h, rebounded off the surface then up and down several times before it could be said to have landed. Luna 9 took the first pictures of the lunar surface and sent them back to Earth.

A year later the American lander Surveyor 3 brought moon dust back to Earth to be analysed. If the dust layer turned out to be too soft, there was the risk that later manned landing modules would sink so deep that they would not be able to take off again – leaving the astronauts stranded. But the mission proved that the dust layer *was* solid enough. Since then more than twenty-five missions have landed on the moon, six of them manned. To date twelve astronauts have walked on the dust-covered surface of the moon – the first to take the famous "giant leap for mankind" being Neil Armstrong and Buzz Aldrin on the American mission Apollo 11 in 1969.

On the basis of the rocks brought back to Earth by Apollo astronauts in the 1970s it has been possible to determine the age of the moon using radioactive dating of different kinds. The age of a rock is defined as its age of crystallisation, i.e. from the point it crystallised from lava mass to solid rock. The time that has elapsed since the rock was lava can be measured using the minerals it contains, minerals locked into the crystal structure of the rock at the end of the lava stage.

The different planets in the solar system have all had lava on the surface at some point. The smaller the planet, the faster it cools down. Its size therefore determines how fast it turns from a liquid ball of lava to develop a hard surface, before finally becoming completely solid. The earth is so large that it still has liquid lava inside, which appears during volcanic eruptions. The small size of the moon – about a third the size of Earth – means that liquid lava no longer exists at its core. The Apollo missions to the moon have shown that volcanic activity ceased on the moon about 1.5 billion years after its formation.

Over the years there have been many interesting theories about the formation and development of the moon. Today the most prevailing hypothesis is that it was created as the result of a massive impact between the earth and another planet. This planet, which has been given the name Theia, is thought to have been the size of Mars. The hypothesis is that Theia collided with a much earlier proto-Earth, hurling large chunks of the earth's surface into a debris disc around the earth. This material later coalesced to form the moon.

The entire solar system was formed around 4.56 billion years ago out of a cloud of gas and dust that collapsed under the force of gravity. The sun formed in the middle, and the planets formed in the disc of debris around the newly formed sun. This explains why all the planets orbit the sun on essentially the same plane, which coincides with the equator of the sun.

The earth was formed by gas and dust on its orbital path around the sun, gathering smaller 'planetesimals' that then collided and coalesced. When the earth was approximately three-quarters assembled – 4.5 billion years ago – it collided with Theia. The massive impact hurled a mass the equivalent of two moons around the earth. Many of the pieces gathered together so close to Earth that tidal forces pulled them back to Earth. The pieces that were thrown far enough away from the earth's gravitational field then joined to form the moon. The pieces that fell back to Earth tilted the orbit of the moon relative to the other moons in the solar system. The entire formation process probably took about a hundred years.

During the early history of the moon the earth spun very fast on its own axis. Day and night probably only lasted a total of five hours, with a few hours of day and an equally short night. The moon was much closer to Earth than it is today, and looked forty times larger than it does now. That the earth and the moon were so close to each other meant that the tidal force was also much more powerful than it is today. This force slowed the rotation of both the moon and the earth, resulting in the moon moving further away from Earth. This still happens today, just on a much smaller scale. Today the moon distances itself from Earth at a rate of around four centimetres a year, and the earth rotates about 1.5 to 2 milliseconds slower per day per century. Over time the earth will slow down so much that it will rotate extremely slowly on its own axis. Ultimately, one side of the earth will constantly face the sun, reaching a state known as tidal locking. This means that one side of the earth will be in constant sunlight, while the other side will lie in constant darkness. A desert will probably develop on the day side, whereas the night side will be covered by ice, possibly with a narrow temperate band between the two extremes.

At the point in the early history of the solar system when the moon was formed, many of the rocks in the solar system had not yet become part of its planets. These rocks bombarded the newly formed earth and moon. The oldest crust of the moon is around 4.5 billion years old, whereas the oldest crust on Earth is around 3.9 billion years old.

Tidal forces have gradually checked the rotation, stopping it completely so the same side always faces Earth. On this side major impacts have created low-lying areas that were filled with basaltic lava during a volcanic period around 3.8 to 3.1 billion years ago. These formed lunar maria that can be seen as large, dark areas on the full moon. On the other side of the moon, facing away from Earth, there are no such low-lying areas. This is because there were fewer free-flying rocks in the solar system compared to when the lunar maria were formed due to a drastic decrease in meteor bombardment.

According to this hypothesis, the moon is primarily made from crust material from Earth and Theia. Most of the iron core of Theia became part of the iron core of the earth. The moon is therefore expected to have a much smaller iron core than the earth. That the average density of the moon is lower than the earth (3.3 g/cm^3 versus 5.5 g/cm^3) supports this hypothesis. Otherwise we would have to assume that lunar rocks are much lighter than rocks on Earth, which is not the case with the c. 380 kg of lunar rocks astronauts have brought back to Earth. Computer simulations of the massive impact between protoplanets show that this hypothesis is feasible. There are, however, still unanswered questions about the size of both the proto-Earth and Theia, and we still do not know how much material from each of these bodies formed the moon.

The finely powdered layer of small minerals on the surface of the moon was formed by meteorite falls. Meteorites are space rocks of different sizes that fly freely in the solar system. Today most of them originate in the asteroid belt between the planets Mars and Jupiter. Asteroids are minor planets that sometimes collide, breaking into smaller fragments. These smaller fragments are often propelled in the direction of the inner solar system – towards the sun. The asteroid fragments are caught by the gravitational field of a planet or moon and deposited on its surface. They are called meteorites once they have landed on the surface, and meteoroids – or asteroids if very large – when flying in space. Space rocks that pass down through the earth's atmosphere are called meteors or fireballs, because the rocks get so hot that their outer layer melts on their way down.

Because the moon has no atmosphere, meteoroids do not heat up or melt in the same way. When they hit the surface of the moon, a crater is formed. The size of the crater depends on the size and speed of the space rock. Meteoroids either crush some of the lunar rocks they hit on the surface, or melt them with the heat generated by the collision, smashing the space rock into a cloud of smaller particles. Over billions of years this process has created the thick layer of dust, sand and gravel seen on the surface of the moon today.

From the moon the sky looks black both during the day and at night, something also due to the lack of atmosphere. If not blinded by the sun, we can also observe stars during the day. Stars in the lunar sky shine with a constant light – they do not 'twinkle' as they appear to do on Earth. It is turbulence in the atmosphere of the earth that 'spreads' stars across a larger area, creating the characteristically shaped star we put on top of the Christmas tree. On the moon stars are well-defined, small point sources.

Solar radiation and electrically charged particles released during solar eruptions hit the surface of the moon unimpeded. A solar eruption is an extreme event in the outer layer of the sun, resulting in the ejection of some of its hot gas. This gas consists of a large number of electrically charged particles. During a solar eruption, these particles are ejected from the sun and can collide with the surface of the moon on their way through space. Since the moon has no atmosphere, the particles hit its surface without being filtered or slowed down as they would be by the atmosphere of Earth. The minerals on the surface of the moon are therefore exposed directly to these solar particles, also called 'space weather'. Space weather is the reason moon dust becomes electrically charged and therefore sticks to the astronauts' spacesuits. When the spacesuit enters the lander, the oxygen and moisture in the air inside results in the particles losing their charge and floating around the cabin.

cat. 5
Buzz Aldrin
The forbidding lunar farside,
Apollo 11, July 1969

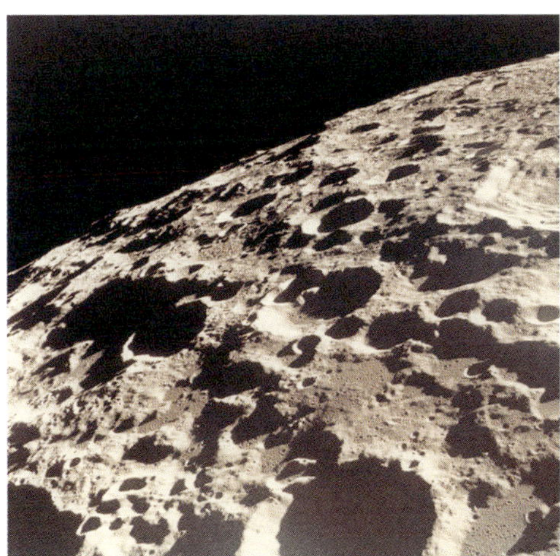

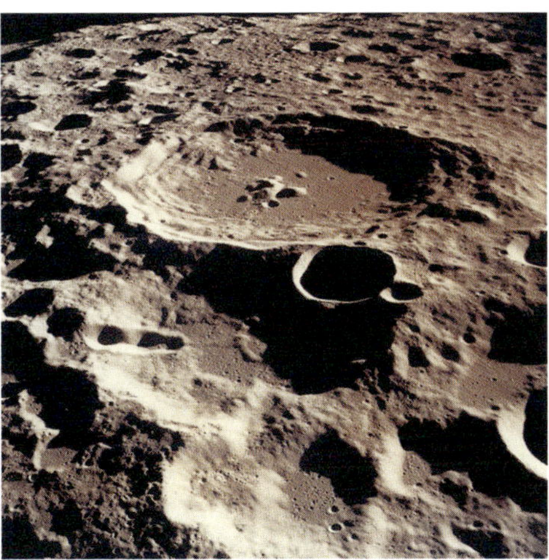

cat. 3 & cat. 4
Buzz Aldrin
The astronaut's boot and foot-print on the Moon, Apollo 11, July 1969

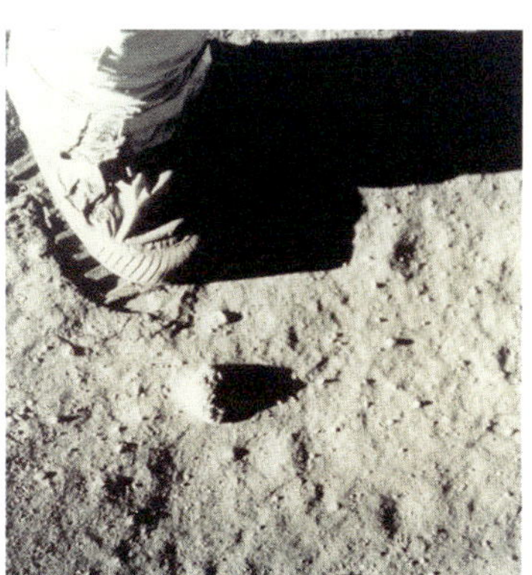

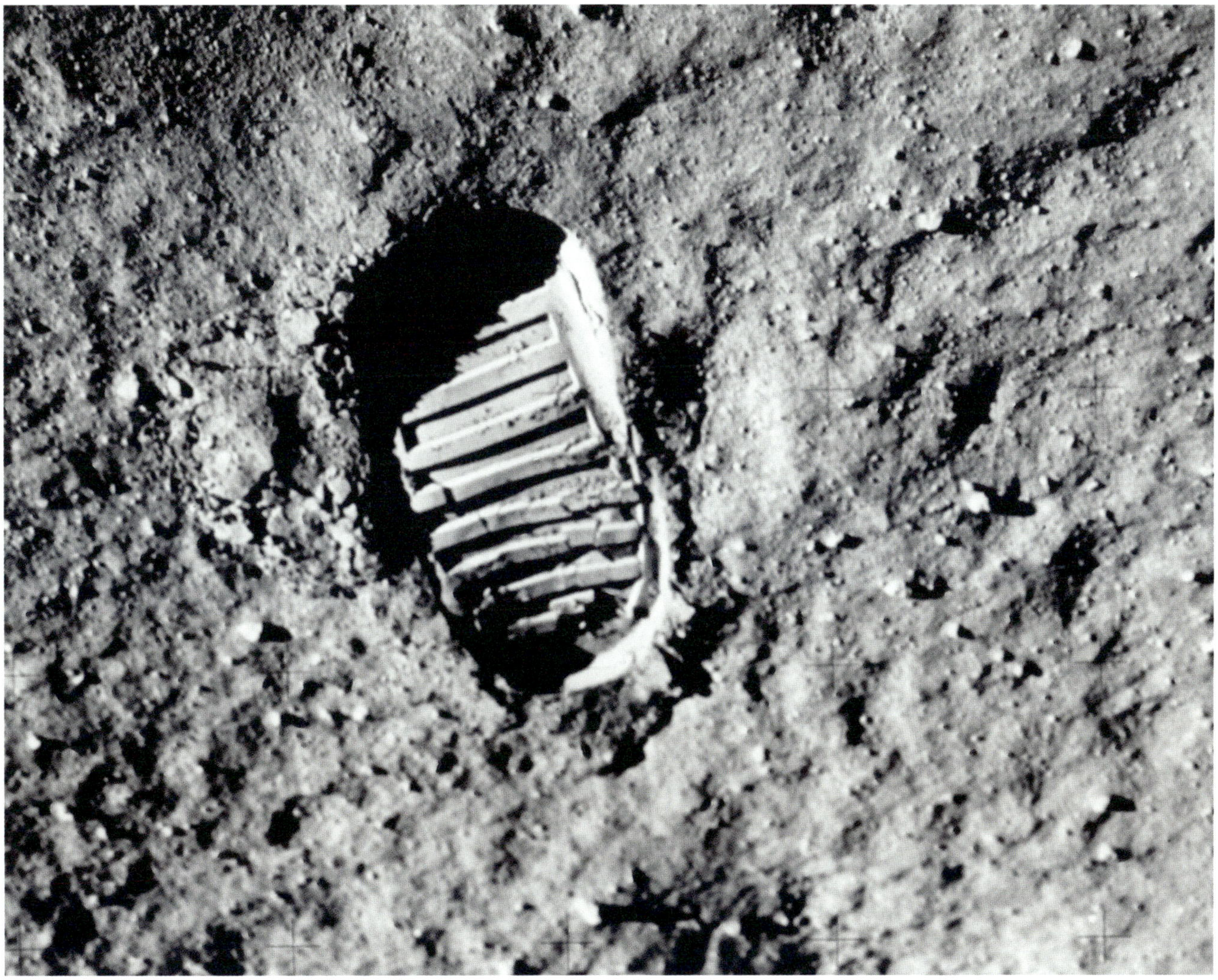

The moon's lack of atmosphere also partially explains why it is so arid: without an atmosphere to create pressure, when ice evaporates it goes directly from ice to vapour without becoming liquid water. If we were to pour water out of a bottle it would evaporate immediately. The earth's atmosphere makes it possible for water to exist as a liquid and not solely as vapour or ice. This is also why we don't see the surface of the moon eroding as it does on Earth in the face of water, wind and volcanoes. The most important change in the surface of the moon over the past many billions of years is due to craters formed by the impact of meteoroids. Craters on the moon therefore tell us a lot about the history of the moon, which can also tell us something about a stage of the history of the earth where we no longer have any evidence due to past meteorite craters being eroded by wind and weather. Many later unmanned missions to the moon have focused on identifying ice deposits in the deep, dark craters on the moon not reached by the rays of the sun, where any ice deposits would therefore be protected from its heat. Finding ice would be essential to a manned base on the moon, since it would provide access to water.

There are around half a million craters with a diameter exceeding one kilometre. There are numerous indications that most of these lunar craters are the result of the 'late heavy bombardment' of the moon and earth 3.8 billion years ago, during which 100 trillion tons of material hit the surface of the moon in a space of around 100,000 years. Later impacts can be seen as craters on top of earlier craters. The footprints of astronauts on the moon will probably be there during the rest of the moon's life, since nothing is erased by wind and weather.

Even though scientists have made major progress since the moon landings – establishing, for example, that lunar matter originally came from Earth – one mystery remains: that the earth is the only rocky planet in the solar system with a moon of significant size. Neither Mercury nor Venus has any moons, and Mars' two small moons, named after the Greek gods Phobos and Deimos ('fear' and 'panic'), are probably captured asteroids from a later stage in the history of Mars. The earth is therefore unique in having such a large moon.

The moon is relatively close to Earth – eleven hours away in a rocket travelling 40,000 km/h. By comparison, it would take half a year to travel to the sun at the same speed. Right now the moon is at an almost magical distance from Earth, covering exactly the same area in the sky as the sun, because the moon is four hundred times smaller than the sun, and the sun is four hundred times further away than the moon. This is what makes it possible to experience total solar eclipses, where the moon totally covers the light of the sun. Not all eclipses are total. Some are annular eclipses, where the moon does not cover the entire surface of the sun so its outer area can be seen as a luminous ring around the moon. That the moon does not always cover the sun entirely is due to the path of the moon around Earth being slightly elliptical rather than circular, so the moon is not always at the same distance from Earth. With time, total solar eclipses will become increasingly rare as the moon moves further and further away from Earth, making

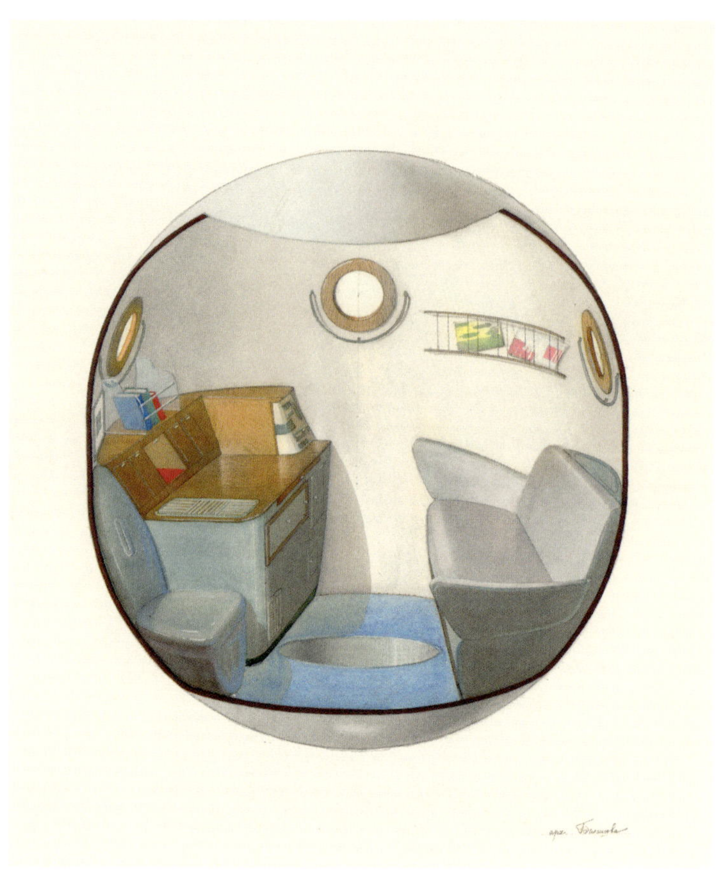

cat. 20
Galina Balashova
Final design for the interior of the Soyuz Orbital Module, approved by Sergey P. Korolev, on 18 February 1964, 1964

cat. 19
Galina Balashova
Initial design for the interior of a Soyuz space capsule, 1963

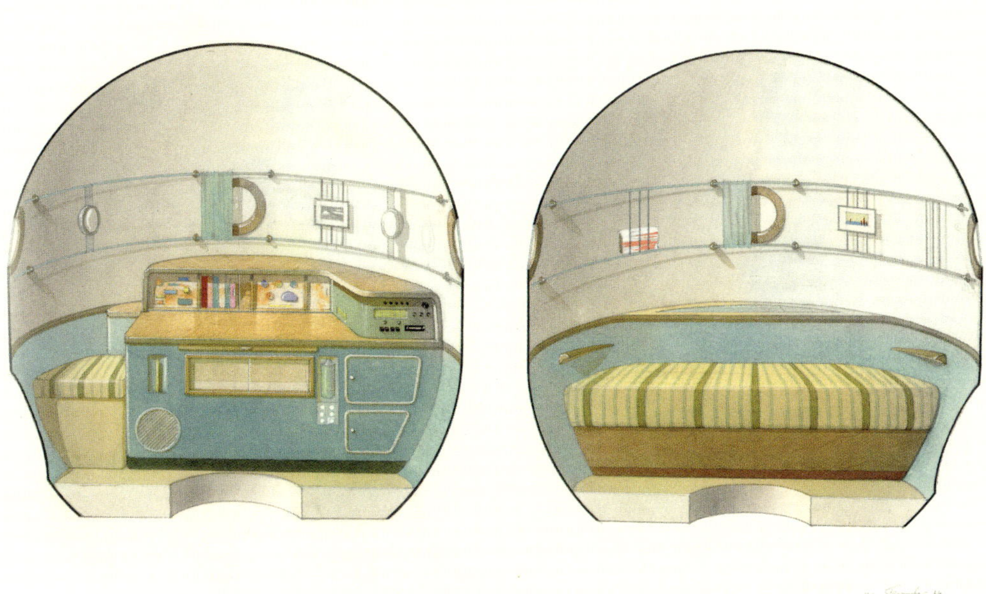

it impossible for it to cover the entire disc of the sun.

It takes around twenty-seven days for the moon to rotate around Earth, and the earth rotates around on own axis once every 24 hours. This is why the moon is sometimes visible at night and sometimes during the day, although we always notice it more at night. The moon is fascinating to behold, both in daylight as an illuminated object against a blue background with only a few visible details, and at night when it stands out in the night sky, especially at full moon. Either as a huge orb close to the horizon, or as a smaller apparently brighter globe high in the sky. That the moon seems larger when it is close to the horizon than high in the sky is an optical illusion. The moon is virtually always the same size. It just seems larger when it is close to the horizon because we can compare it to the size of something we know, whereas when it is high in the sky we can only compare it to the stars close by.

The moon is larger than the dwarf planet Pluto, which is at the edge of the solar system. It is actually not the size of an object that determines whether we call it a planet or a moon, but the orbit it is in. If an object orbits the sun (or another star) directly we call it a planet, regardless of size. We do, however, distinguish between gas planets, earth-like planets, dwarf planets and minor planets. If an object orbits a planet we call it a moon, again regardless of size. Jupiter's moon Ganymede and Saturn's moon Titan are both larger than Pluto and the planet Mercury. Our moon is the fifth largest moon in the solar system, which coincidentally fits with the earth being the fifth largest planet in the solar system.

The light of the full moon is strong enough to read by, and it outshines the starry sky it lies in. For an astronomer it is therefore always better to use a telescope during a new moon than a full moon, since the light of the moon can easily swamp the stars and galaxies you are trying to observe. But when the moon is out, it is always worth finding a pair of binoculars to look at this orb that is so close and yet so incredibly far away. With binoculars we can see the structures on the moon – its craters, peaks and valleys. It is easier to see details on the moon with binoculars in the period between a new moon and a half moon when the moon is not too bright. But the best view of the moon is during a lunar eclipse, because the shadow of the earth reduces the brightness of the moon, allowing more detail to emerge. Then we can stand here on Earth in the reddish glow of the eclipsed moon and dream of being up there as a space tourist, imagining what it would be like to walk between the craters and leave our own permanent footprints in the powdery dust of the moon.

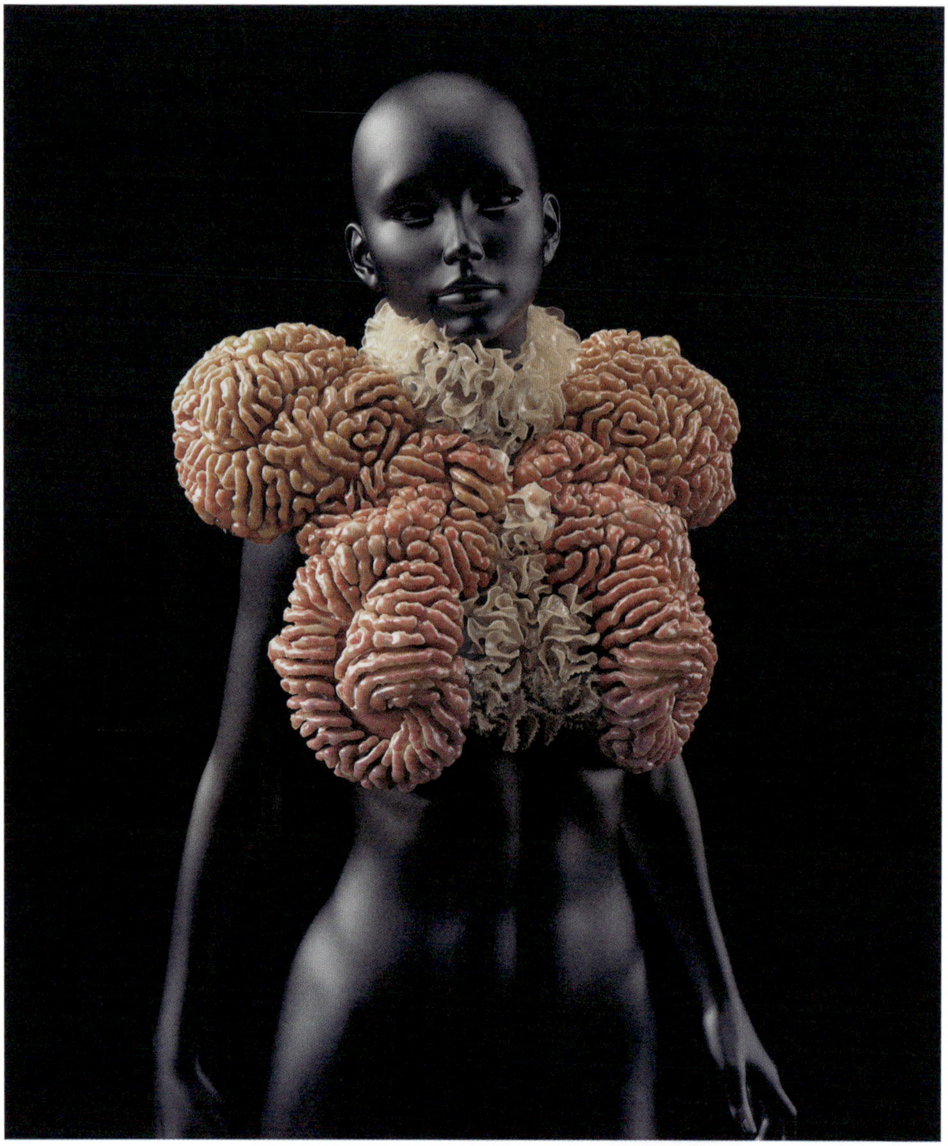

cat. 101
Neri Oxman
Qmar – Luna's Wanderer, 2014
Detail below

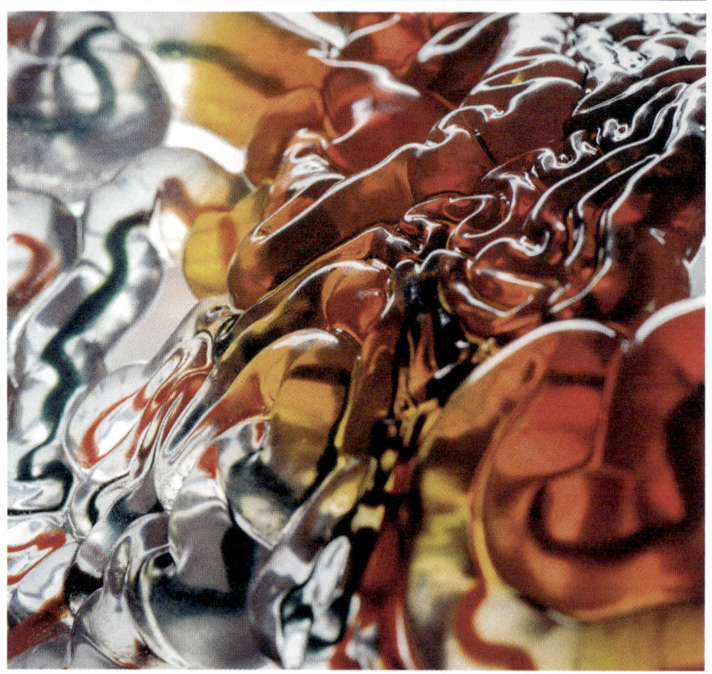

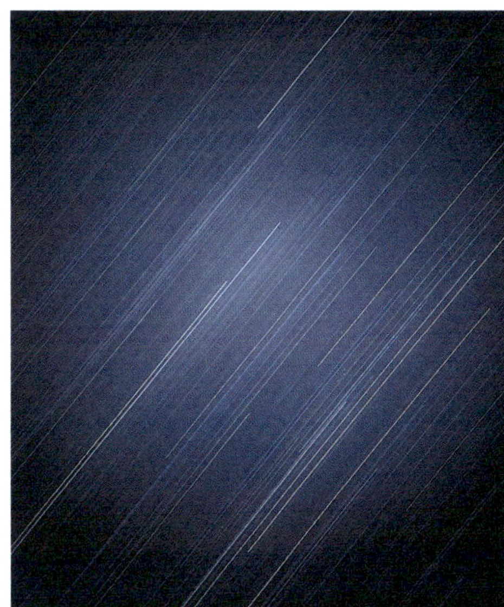

cat. 106
Trevor Paglen
PAN (Unknown; USA-207), 2010-11

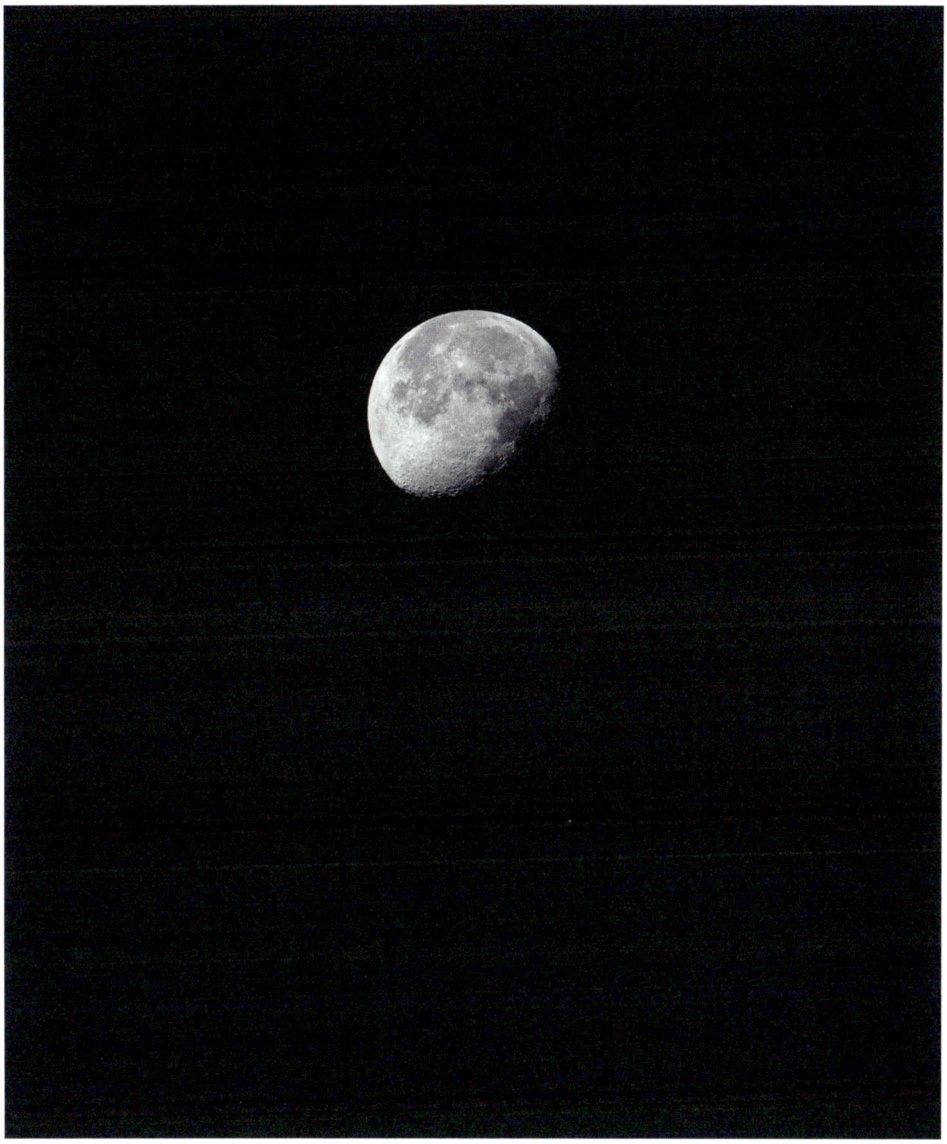

cat. 105
Trevor Paglen
Dead Military Navigation Satellite (COSMOS 985) Near the Disk of the Moon, 2012

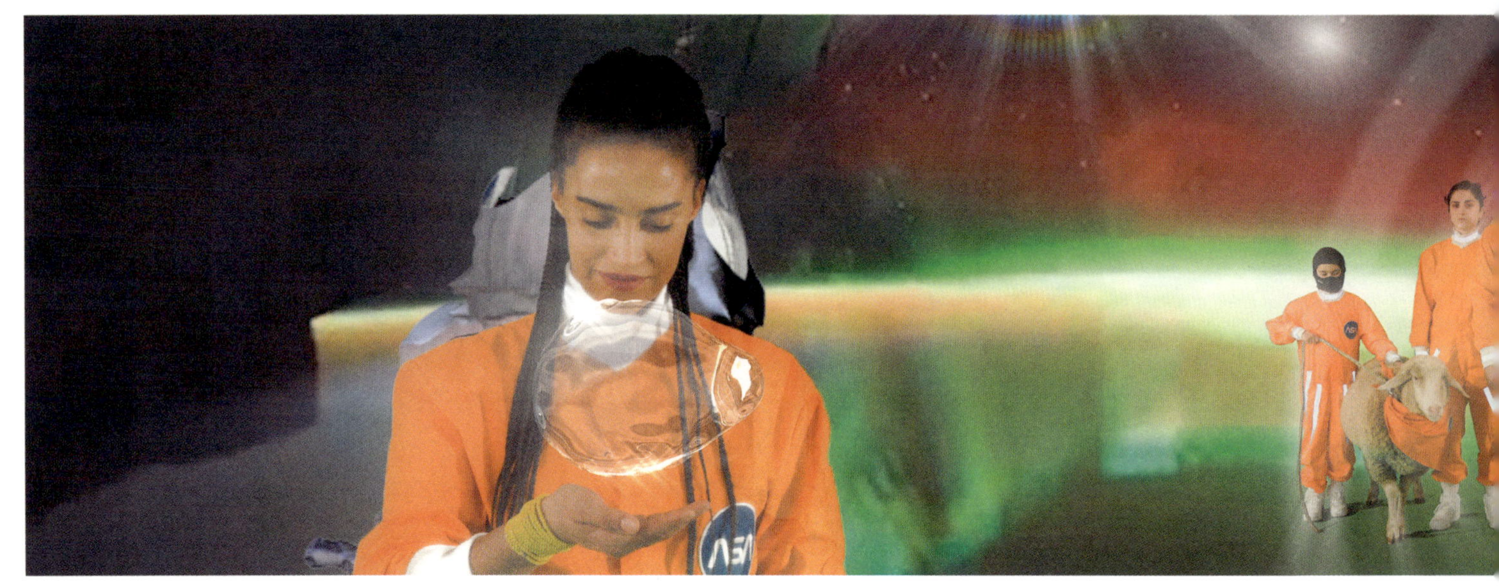

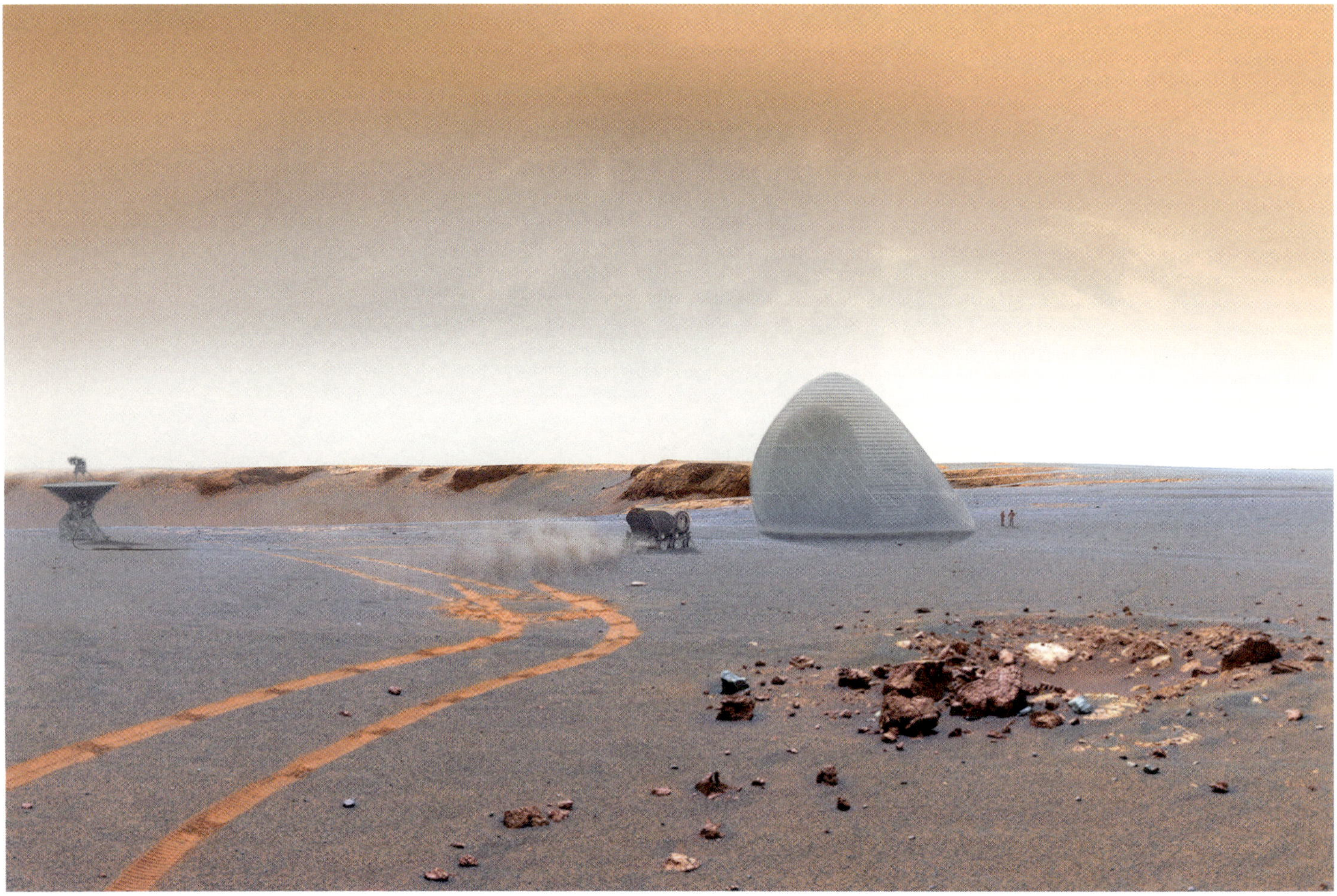

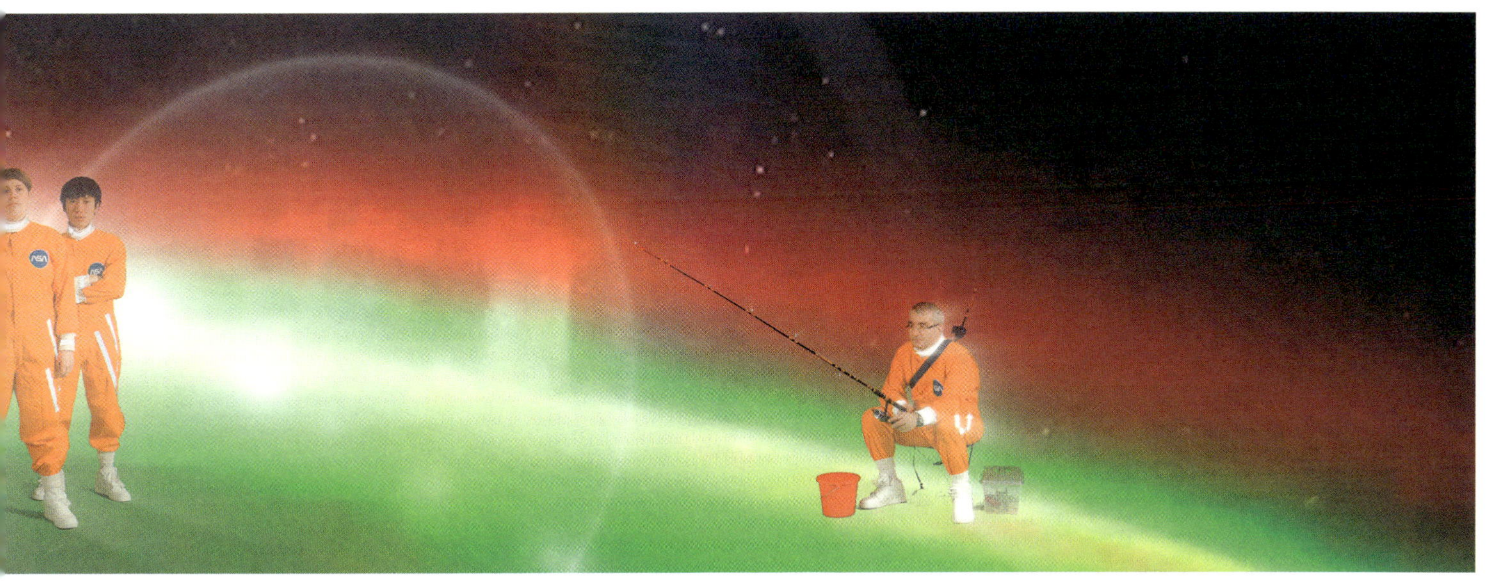

cat. 163
Hito Steyerl
ExtraSpaceCraft, 2016

cat. 157
SEArch & CLOUDS
Architecture Office
Mars Ice House, 2015

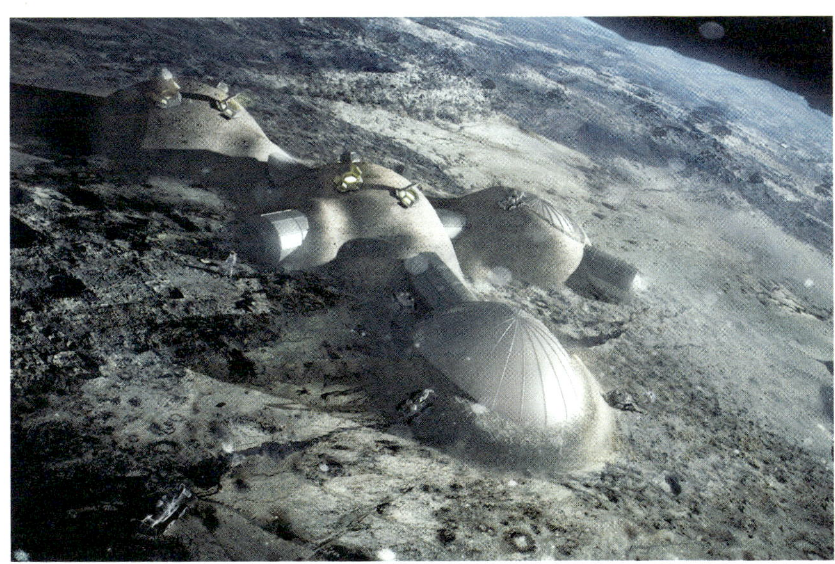

cat. 57
Foster + Partners
Lunar Settlements, 2012

List of Works

1 **Gertrude Abercrombie**
The Courtship, 1949
Oil on Masonite panel, 55.2 × 64.1 cm
Collection Museum of Contemporary Art Chicago, gift of the Gertrude Abercrombie Trust, 1978.56

2 **Muhammad ibn Ahmad al-Mizzi**
Astro Lab Quadrant, 1329-30
Brass, engraved, H: 15.5 cm, D: 1.3 cm
The David Collection, Copenhagen
(Inv.No. 16/1988)

3 **Buzz Aldrin**
The astronaut's footprint on the Moon, Apollo 11, July 1969
Vintage gelatin silver print, 20.3 × 25.4 cm
Collection Victor Martin-Malburet

4 **Buzz Aldrin**
Aldrin's boot in lunar soil, Apollo 11, July 1969
Vintage chromogenic print on fiber-based Kodak paper, 20.3 × 25.4 cm
Collection Victor Martin-Malburet

5 **Buzz Aldrin**
The forbidding lunar farside, Apollo 11, July 1969
Diptych: Two vintage chromogenic prints on fiber-based Kodak paper, à 20.2 × 25.4 cm
Collection Victor Martin-Malburet

6 **Buzz Aldrin**
The LM "Eagle" on the Moon, Apollo 11, July 1969
Vintage chromogenic print on resin-coated Kodak paper, 20.3 × 20.3 cm
Collection Victor Martin-Malburet

7 **Darren Almond**
Fullmoon@Fell, 2005
C-print, 127.5 × 127.5 cm, framed
Courtesy the Artist and White Cube

8 **Darren Almond**
Fullmoon@Gordale Scar, 2005
C-print, 127.5 × 127.5 cm, framed
Courtesy the Artist and White Cube

9 **Darren Almond**
Fullmoon@Peat Brook, 2007
C-print, 127.5 × 127.5 cm, framed
Courtesy the Artist and White Cube

10 **Darren Almond**
Fullmoon@Wester Ross, 2007
C-print, 127.5 × 127.5 cm, framed
Courtesy the Artist and White Cube

11 **Darren Almond**
Fullmoon@Yesnaby, 2007
C-print, 127.5 × 127.5 cm, framed
Courtesy the Artist and White Cube

12 **Darren Almond**
Fullmoon@Baltic Coastline, 2015
Latex print, 188.2 × 400 × 5.9 cm
Courtesy of the Artist and White Cube

13 **Abd al-Rahman al-Sufi**
Kitab suwar al-kawakib, 1675
Book of Fixed Stars
Paper, polychrome ink, gold and silver, page: 24.5 × 15 cm, open with flap: 24.8 × 44.3 cm
The David Collection, Copenhagen
(Inv.No. 165/2006)

14 **William Anders**
First photograph of the planet Earth taken by an astronaut, Apollo 8, December 1968
Vintage chromogenic print on fiber-based Kodak paper, 20.3 × 25.4 cm
Collection Victor Martin-Malburet

15 **Laurie Anderson & Hsin-Chien Huang**
Moon, September 2018
Virtual reality and physical set
Laurie Anderson & Hsin-Chien Huang

16 **Neil Armstrong**
Buzz Aldrin about to take his first step on the Moon, Apollo 11, July 1969
Vintage chromogenic print on fiber-based Kodak paper, 20.3 × 25.4 cm
Collection Victor Martin-Malburet

17 **Neil Armstrong**
Buzz Aldrin's gold-plated sun visor reflects the photographer and the Lunar Module, Apollo 11, July 1969
Vintage chromogenic print on resin-coated Kodak paper, 20.3 × 25.4 cm
Collection Victor Martin-Malburet

18 **Neil Armstrong**
First photograph of a man standing on the surface of another world, Apollo 11, July 1969
Vintage chromogenic print on fiber-based Kodak paper, 20.3 × 25.4 cm
Collection Victor Martin-Malburet

19 **Galina Balashova**
Initial design for the interior of a Soyuz space capsule, 1963
Watercolour on paper, 61.7 × 44.7 cm

20 **Galina Balashova**
Final design for the interior of the Soyuz Orbital Module, approved by Sergey P. Korolev, on 18 February 1964, 1964
Watercolour on paper, 50.7 × 78.7 cm

21 **Galina Balashova**
Initial sketches of the interior of LOK, 1966
Watercolour on paper, 43.8 × 82.8 cm

22 **Galina Balashova**
Initial sketches of the interior of LOK, 1966
Watercolour on paper, 38 × 74 cm

23 **Galina Balashova**
Design for the interior of LOK, portside, 1967-1968
Watercolour on paper, 79 × 58 cm

24 **Rosa Barba**
The Color Out of Space, 2015
HD video, colour, sound; 5 coloured glass filters, steel base 36 min. 125 × 112 × 56 cm
Courtesy of the Artist

25 **Wilhelm Beer**
Mappa selenographica totam lunare hemispaeram visibilem complectens, 1834
Book, 54.2 × 55.8 cm (page),
108.5 × 111.5 cm (open, lying down)
Sächsische Landesbibliothek – Staats- und Universitätsbibliothek Dresden

26 **Henry Blunt**
Electrotype copy of a plaster model showing the lunar crater; Eratosthenes in a hinged mahogany case, 1849-1851
Metal, mahogany and paper, 16.5 × 25.5 cm
Science Museum Group, London

27 **Tycho Brahe**
Tychonis Brahei Ottonidis Genethliacon Inclyto Infanti Huldarico, Fred. IIdi Filio Secundo Genito, Dicatum 1579, 1579
Book, 14.8 × 21 cm
The Royal Danish Library, Copenhagen

28 **Tycho Brahe**
Tychonis Brahei Volumen [III] Observationum Pro Annis 1590-94 Incl., 1590
Book, 21 × 29.7 cm
The Royal Danish Library, Copenhagen

29 **Nanna Debois Buhl**
Moon Memory, 2018
Series of prints, photogravure
1 print: 87 × 64.7 cm / 6 prints: 87 × 117 cm
Courtesy of the Artist

30 **Charles Conrad**
Unidentified Flying Object, Gemini 11, September 1966
Vintage chromogenic print on fiber-based Kodak paper, 20.3 × 25.4 cm
Collection Victor Martin-Malburet

31 **Joseph Cornell**
Le Voyageur dans les Glaces, Jouet Surréaliste, 1935
The Traveller on Ice, Surrealist Game
Box / Game comprising 7 discs and a device to spin the discs, 9.3 × 9.3 × 2.8 cm
Mark Kelman, New York

32 **Joseph Cornell**
Soap Bubble Set, 1948
Box construction, 36.8 × 52.1 × 9.8 cm
Mr. & Mrs. John Stravinsky

33 **Joseph Cornell**
Sun Box, 1960
Assemblage, 25.7 × 38.7 × 8.9 cm
Whitney Museum of American Art, New York
Gift of Howard and Jean Lipman

34 **Joseph Cornell**
Untitled (Compartmented Box), 1954-1956
Box construction, 38.7 × 26 × 6.2 cm
Moderna Museet, Stockholm

35 **Joseph Cornell**
For Angela (Manila paper unevenly stained blue), 1969-1970
Collage, 22.9 × 17.8 cm
Smithsonian American Art Museum,
Gift of The Joseph and Robert Cornell Memorial Foundation

36 **Joseph Cornell**
Dovecote, c. 1954
Assemblage, 27 × 14.8 × 9.5 cm
Mark Kelman, New York

37 **Joseph Cornell**
Untitled (Solar Set), c. 1956-58
Assemblage, 29.2 × 41.3 × 9.2 cm
Collection of Robert Lehrman,
Courtesy of Aimee and Robert Lehrman

38 **Joseph Cornell**
"Ideals are like stars; you will not succeed in touching them with your hands. But like the seafaring man on the desert of waters, you choose them as your guides, and following them you will reach your destiny." – Carl Schurz, Address, Faneuil Hall, Boston, April 18, 1859. From the series Great Ideas of Western Man, c. 1957-1958
Mixed media: painted and stained wood, glass, shells, driftwood, and paper, assembled, glued, and nailed
44.2 × 32.7 × 8.9 cm
Smithsonian American Art Museum,
Gift of Container Corporation of America

39 **Donato Creti**
Le osservazioni astronomiche, 1711
Astronomical Observations
Oil on canvas, 8 panels, each 51 × 35 cm
Vatican Museums, Vatican City

40 **Walter Cunningham**
The Sun illuminating the Earth above the Florida Peninsula, Apollo 7, October 1968
Vintage chromogenic print on fiber-based Kodak paper, 20.3 × 25.4 cm
Collection Victor Martin-Malburet

41 **J.C. Dahl**
Bugten ved Napoli, 1821
The Bay at Naples
Oil on canvas, 49.7 × 68 cm
Thorvaldsens Museum, Copenhagen

42 **J.C. Dahl**
Bugten ved Napoli set fra en grotte, 1821
The Bay of Naples seen from a Grotto
Oil on canvas, 22.2 × 33.4 cm
Thorvaldsens Museum, Copenhagen

43 **J.C. Dahl**
Måneskin over havet, c. 1820-35
Moonlight over the Sea
Oil on canvas on cardboard, 14.5 × 21.5 cm
SMK, National Gallery of Denmark

44 **Salvador Dalí**
Girl with Curls, 1926
Oil on panel, 50.8 × 40 cm
Collection of The Dalí Museum,
St. Petersburg, FL

45 **Salvador Dalí**
Big Thumb, Beach, Moon and Decaying Bird, 1928
Oil, sand and gravel on panel, 50.2 × 91 cm
Collection of The Dalí Museum,
St. Petersburg, FL

46 **Charles Delagrave**
European Celestial Globe on wooden pillar stand, 1878
Oak, wood (unidentified), paper, plaster and brass, 185 × diam. 107 cm
Science Museum Group, London

47 **John W. Draper**
First known photograph of the Moon was taken by John W. Draper c. 1839-40. The spots in this photo are caused by mold and water damage on the original daguerreotype, which apparently no longer exists
Photography
New York University Archives

48 **John W. Draper**
This daguerreotype has been attributed to John Draper from the period in the 1840s when he was creating this type of moon image
Photography, 8.25 × 7 × 0.9 cm
New York University Archives

49 **Charles Duke**
John Young jumps and salutes the flag, EVA 1, Apollo 16, April 1972
Vintage chromogenic print on fiber-based Kodak paper, borderless, 20.3 × 25.4 cm
Collection Victor Martin-Malburet

50 **Charles Duke**
Panoramic view of the Descartes landing site with the LM "Orion", John Young, the American flag and the solar wind, EVA 2, Apollo 16, April 1972
Mosaic of four vintage chromogenic prints, 24 × 68 cm
Collection Victor Martin-Malburet

51 **Marlene Dumas**
Skyros, 2016
Oil on panel, diam. 150 cm
Courtesy the Artist and David Zwirner

52 **Charles Eames**
A Rough Sketch for a Proposed Film Dealing with the Powers of Ten and the Relative Size of Things in the Universe, 1968
Film, 8 min, loop
a film by Charles and Ray Eames, 1968 © 2017 Eames Office, LLC (eamesoffice.com)

53 **C.W. Eckersberg**
Måneskinsbillede, 1821
Moonlight Painting
Oil on canvas, 48 × 63.5 cm
The Nivaagaard Collection

54 **Max Ernst**
Moonmad, 1944
Bronze, 92.6 × 32.1 × 29.8 cm
Fondation Beyeler, Riehen/Basel
Beyeler Collection

55 **Max Ernst**
The Twentieth Century, 1955
Oil on canvas, 50.8 × 61 cm
Max Ernst Museum Brühl des LVR,
Schenkung Dr. Peter Schneppenheim

56 **Max Ernst**
Naissance d'une galaxie, 1969
The Birth of a Galaxy
Oil on canvas, 92 × 73 cm
Fondation Beyeler, Riehen/Basel
Beyeler Collection

57 **Foster + Partners**
Lunar Settlements, 2012
Timber, nylon, acrylic and 3D printed, 50 × 92 × 182 cm
Foster+Partners

58 **Caspar David Friedrich**
Meeresküste im Mondlicht, 1818
Seashore by Moonlight
Oil on canvas, 22 × 36.7 cm
Paris, Louvre Museum, Department XXX

59 **Galileo Galilei**
Astronomia, manoscritto cartaceo, autografo, Galileiano 48
November – December 1609
Watercolour drawing of the phases of the Moon from the autograph manuscript redaction of Sidereus Nuncius
Manuscript
33 × 23 × 1.7 cm
Biblioteca Nazionale Centrale Firenze

60 **Galileo Galilei**
Sidereus Nuncius, 1610
Starry Messenger
Book, 24.5 × 18 cm
The Royal Danish Library, Copenhagen

61 **Leopold Galluzzo**
Altre scoverte fatte nella luna dal Sigr. Herschel, 1836
Lithograph, hand coloured, 6 leafs, each 50.8 × 61 cm
Smithsonian Libraries, Washington DC

62 **Isa Genzken**
Wind (D), 2009
Wood, fabric, ceramics, colour prints on paper, tape, spray paint, CD's and metal clips, 315 × 190 × 110 cm
Galerie Buchholz, Cologne/Berlin/New York

63 **Isa Genzken**
Vollmond, 2015 (Model)
Full Moon
Plastic, metal, acrylic paint, wood and LED, 150 × 50 × 95 cm
1:75
Galerie Buchholz, Cologne/Berlin/New York

64 **Richard Hamilton**
Towards a definitive statement on the coming trends in menswear and accessories (a)
Together let us explore the stars, 1962
Oil paint, cellulose paint and printed paper on wood, 61 × 81.3 cm
Tate: Purchased 1964

65 **Camille Henrot**
October 2015 Horoscope, 2015
Resin, ardiuno motor, LED strobe lights, splitter connectiong LED and motor, control swicth, 96.5 × 223.5 × 254 cm
Courtesy the Artist and Metro Pictures, NY

66 **Johannes Hevelius**
Selenographia, sive Lunae descriptio, 1647
Selenography, or A Description of The Moon
Book, H: 37 cm
The Royal Danish Library, Copenhagen

67 **James Irwin**
Panorama of Hadley Rille lunar canyon and the Apennine Mountains, Station 10, EVA 3, Apollo 15, August 1971
Mosaic of seven vintage gelatin prints numbered in black in top margin, 45 × 114 cm
Collection Victor Martin-Malburet

68 **James Irwin**
Panoramic view with David Scott and the lunar rover on the edge of Hadley Rille Canyon, Station 2, EVA 1, Apollo 15, August 1971
Mosaic of five vintage gelatin silver prints numbered in black in top margin, 25 × 90 cm
Collection Victor Martin-Malburet

69 **Marie Kølbæk Iversen**
Moonologue. En opera for to sangere, 2018
Moonologue. An Opera for Two Singers
Composition by Katinka Fogh Vindelev
Lyrics and idea by Marie Kølbæk Iversen
Performance, Louisiana 11.23. & 12.7. 2018

70 **Marie Kølbæk Iversen**
IO / I, 2015-
Digital 3D animations of NASA footage, Variable dimensions
Courtesy Marie Kølbæk Iversen & Gether Contemporary

71 **Athanasius Kircher**
Ars Magna Lucis Et Umbrae, 1646
The Great Art of Light and Shadow
31 × 40 cm
The Royal Danish Library, Copenhagen

72 **Yves Klein**
Pink Planetary Relief "Lune II" (RP 21), 1961
Dry pigment and undefined binder on plaster, 95 × 65 cm
Private collection

73 **Yves Klein**
Planetary Relief "Region de Grenoble" (RP 10), 1961 (Posthumous edition c. 1990)
Dry pigment and synthetic resin on bronze, 86 × 65 cm
Private collection

74 **Yves Klein**
Untitled White Planetary Relief (RP 12), c. 1961
Dry pigment and undefined binder on plaster on panel, 96 × 69 cm
Private collection

75 **Yves Klein with the collaboration of Harry Shunk and John Kender**
Leap into the Void, 1961 (org.) – 2018 (exhibition print)
Photograph, 150 × 190 cm
Private collection

76 **Kiki Kogelnik**
Fly Me to the Moon, 1963
Oil and acrylic on paper, 244.2 × 184.2 cm
Courtesy of Mono Schwarz-Kogelnik

77 **Kiki Kogelnik**
Outer Space, 1964
Oil and acrylic on canvas, 182.9 × 137.2 cm
Courtesy of Kiki Kogelnik Foundation

78 **Sergei Pavlovich Korolov (Designer) & Oleg Ivanovsky (Designer)**
Sputnik 1 satellite (Replica), 1974
Metal, 55 × 55 × 350 cm
Science Museum Group, London

79 **Neil Armstrong's Spacesuit from Apollo 11 (replica)**
Spaceconsult

80 **Alicja Kwade**
Freiheitsgrad, 2017
Degree of Freedom
Stainless steel, stone, 185 × 206 × 210 cm
Courtesy of the Artist and KÖNIG GALERIE, Berlin / London

81 **Alicja Kwade**
Pars pro Toto, 2018
8 stone globes perfectly round, Size varies: 70 – 150 cm
Courtesy of the Artist and KÖNIG GALERIE, Berlin / London

82 **Alicja Kwade**
REVOLUTION (Gravitas), 2018
Stainless steel, stone, c. 315 × 320 × 285 cm
Courtesy of the Artist and KÖNIG GALERIE, Berlin / London

83 **Michael Florent van Langren**
Plenilunii Lumina Austriaca Philippica, 1645
The Lights of Philip of Austria
Etching, 38 × 50 cm 60 × 60 × 2 cm
Collection Bibliothèque Nationale et Universitaire de Strasbourg

84 **Cath Le Couteur**
SuitSat, Vanguard and Fengyun, 2017
Director: Cath Le Couteur
Voiced by Gruff Rhus, Sally Potter & Delilah Holiday
Film, 6 min, loop
Courtesy the Artist
projectadrift.co.uk

85 **Carl Julius Leypold**
Kirchhofseingang, 1832
Entrance to the Cemetery
Oil on canvas, 25.5 × 35 cm
Germanishes Nationalmuseum, Nürnberg
Loan by the city of Nürnberg

86 **Michael Madsen & Jonathan Houser**
The Search, 2018
Interactive installation, 300 × 750 × 750 cm
Courtesy of the Artists

87 **James McDivitt**
First US Spacewalk – Ed White's EVA (Extra Vehicular Activity), Gemini 4, 3 June 1965
Vintage chromogenic print on fiber-based Kodak paper, 20.3 × 25.4 cm
Collection Victor Martin-Malburet

88 **James McDivitt**
First US Spacewalk – Ed White's EVA over New Mexico, Gemini 4, 3 June 1965
Vintage chromogenic print on fiber-based Kodak paper, 20.3 × 25.4 cm
Collection Victor Martin-Malburet

89 **James McDivitt**
First US Spacewalk – Ed White's EVA over South California, Gemini 4, 3 June 1965
Vintage chromogenic print on fiber-based Kodak paper, 20.3 × 25.4 cm
Collection Victor Martin-Malburet

90 **Georges Méliès**
La Lune à un mètre, 1898
An Astronomer's Dream
Film, 3 min, 11 sec, loop
Lobster Films Collection

91 **Georges Méliès**
Le Voyage dans la Lune, 1902
A Trip to the Moon
Film, hand-coloured, 14 min, loop
Crédits Lobster-Fondation Groupama Gan-Fondation Technicolor, Courtesy of mk2 Films

92 **Georges Méliès**
Éclipse du Soleil en pleine Lune, 1907
The Eclipse
Film, 9 min, loop
Lobster Films Collection

93 **Claude Mellan**
Trois phases de la lune, 1636
Three Phases of the Moon
Etching, 55.3 × 40.6 cm
Abbeville, Musée Boucher De Perthes

94 **Edvard Munch**
Måneskinn ved stranden, 1893
Moonlight at the Beach
Oil on canvas, 62.5 × 95.8 cm
KODE kunstmuseer og komponisthjem

95 **James Nasmyth**
Plaster relief model, of a portion of the Moon's surface, 1850-1871
Wood and plaster, 5.5 × 47.5 × 58 cm
Science Museum Group, London

96 **James Nasmyth**
The Moon: Considered as a Planet, a World, and a Satellite, London, 1874
Book, 28.7 × 22.7 cm
The Royal Danish Library, Copenhagen

97 **James Arthur O'Connor**
The Poachers, 1835
Oil on canvas, 55.5 × 70.5 cm
On Loan from the National Gallery of Ireland, Dublin

98 **Yasuaki Onishi**
Reverse of Volume LM, 2018
Glue, plastic sheet, 200 × 210 × 475 cm

99 **Neri Oxman**
Mushtari – Jupiter's Wanderer, 2014
3D printed on a Stratasys Objet500™ Connex3 Color Multi-material 3D Production System
Materials: VeroClear, VeroMagenta, VeroCyan, 45 × 45 × 40 cm
Design by Neri Oxman in collaboration with Dominik Kolb, Christoph Baden & Joe Hicklin. Art Collection Stratasys Ltd. 3D printed by Stratasys

100 **Neri Oxman**
Otaared – Mercury's Wanderer, 2014
3D printed on a Stratasys Objet500™ Connex3 Color Multi-material 3D Production System
Materials: VeroClear, VeroYellow, VeroCyan, 50 × 35 × 40 cm
Design by Neri Oxman in collaboration with Christoph Baden, Dominik Kolb and Joe Hecklin. Art Collection by Stratasys Ltd. 3D pinted by Stratasys

101 **Neri Oxman**
Qamar – Luna's Wanderer, 2014
3D printed on a Stratasys Objet500™ Connex3 Color Multi-material 3D Production System
Materials: VeroClear, VeroYellow, Vero Magenta, 50 × 40 × 40 cm
Design by Neri Oxman in collaboration Christoph Bader, Dominik Kolb and Joe Hicklin, in partnership with STRATASYS Ltd.

102 **Neri Oxman**
Zuhal – Saturn's Wanderer, 2014
3D printed on a Stratasys Objet500™ Connex3 Color Multi-material 3D Production System
Materials: VeroClear, VeroYellow, VeroCyan, 45 × 50 × 40 cm
Design by Neri Oxman in collaboration with Christoph Baden, Dominik Kolb & Joe Hicklin. Art Collection by Stratasys Ltd. 3D print by Stratasys

104 **Trevor Paglen**
MILSTAR 3 in Sagittarius (Inactive Communication and Targeting Satellite; USA 143), 2008
C-print, 95.3 × 76.2 cm
Courtesy the Artist and Metro Pictures, New York

104 **Trevor Paglen**
KEYHOLE 12-3 (IMPROVED CRYSTAL) Optical Reconnaissance Satellite Near Scorpio (USA 129), 2007
C-print, 149.9 × 120.7 cm
Courtesy the Artist and Metro Pictures, New York

105 **Trevor Paglen**
Dead Military Navigation Satellite (COSMOS 985) Near the Disk of the Moon, 2012
C-print, 111.8 × 91.4 cm
Courtesy the Artist and Metro Pictures, New York

106 **Trevor Paglen**
PAN (Unknown; USA-207), 2010-11
C-print, 152.4 × 121.9 cm
Courtesy the Artist and Metro Pictures, New York

107 **Katie Paterson**
Earth – Moon – Earth (Moonlight Sonata Reflected from the Surface of the Moon), 2007
Yamaha Disklavier Baby-Grand Piano, MIDI file, 100 × 146 × 149 cm
5 min, 54 sec
Courtesy the Artist

108 **Katie Paterson**
Light Bulb to Simulate Moonlight, 2008
289 light bulbs with halogen filaments, frosted coloured shells (28W, 4500K), crate and logbook
Dimensions variable
Courtesy the Artist

109 **Victor Prouvé**
Amour de lune, 1884
Love for the Moon
Pen and ink on paper, 29.5 × 15 cm
Georgina Kelman

110 **Wolfgang Paalen**
Cadran Lunaire, 1935
Disk of the Moon
Oil and tempera on canvas, 130.8 × 83.8 cm
Gordon Onslow Ford Collection, Lucid Art Foundation Courtesy of Gallery Wendi Norris

111 **Wolfgang Paalen**
Untitled, 1940
Oil and fumage on canvas, 40.6 × 40.6 cm
Private Collection, Courtesy of Gallery Wendi Norris, San Francisco

112 **Robert Rauschenberg**
Horn (Stoned Moon), 1969
Lithograph, 112.4 × 86.4 cm
Robert Rauschenberg Foundation

113 **Robert Rauschenberg**
Sky Garden, 1969
Lithograph and silk screen print on paper, 105.5 × 224.5 cm
Louisiana Museum of Modern Art. Long-term loan: Museumsfonden af 7. december 1966

114 **Robert Rauschenberg**
Stoned Moon Drawing, 1969
Collage and crayon on illustration board, 50.5 × 73 cm
Robert Rauschenberg Foundation

115 **Robert Rauschenberg**
Ape (Stoned Moon), 1970
Lithograph, 116.8 × 83.8 cm
Robert Rauschenberg Foundation

116 **Robert Rauschenberg**
Arena II, State II (Stoned Moon), 1970
Lithograph on paper, 119.5 × 81.5 cm
Robert Rauschenberg Foundation

117-
119 **Robert Rauschenberg**
Drawing for Stoned Moon Book, 1970
Collage of photographs, watercolour, coloured pencil and graphite on illustration board, 40.6 × 51.4 cm 50 × 65.7 × 3.2 cm
Robert Rauschenberg Foundation

120-
123 **Robert Rauschenberg**
Drawing for Stoned Moon Book, 1970
Solvent transfer with printed reproductions, photographs, and graphite on illustration board
25.4 × 38.1 cm
Robert Rauschenberg Foundation

124 **Robert Rauschenberg**
Hybrid (Stoned Moon), 1970
Lithograph, 138 × 91.5 cm
Robert Rauschenberg Foundation

125 **Robert Rauschenberg**
Local Means (Stoned Moon), 1970
Lithograph, 82 × 110 cm
Robert Rauschenberg Foundation

126-
137 **Robert Rauschenberg**
Stoned Moon Book, 1970
Collage of photographs, watercolour, press type, acetate, graphite and coloured pencil on illustration board
Variable dimensions
Robert Rauschenberg Foundation

138 Robert Rauschenberg
White Walk (Stoned Moon), 1970
Lithograph, 107.3 × 74.9 cm
Robert Rauschenberg Foundation

139 Man Ray
Le Monde, 1931
The World
Photogravure, 26 × 20.5 cm
Smithsonian American Art Museum,
Museum purchase

140 Giovanni Battista Riccioli
Almagestum Novum, 1653
New Almagest
Book, 35 × 45 cm
The Royal Danish Library, Copenhagen

141 Rachel Rose
Everything and More, 2015
HD video, colour, sound with mylar,
PVC and carpet, 11 min, 33 sec, loop
Courtesy of Pilar Corrias

142 Mark Rothko
Untitled, 1969
Acrylic on canvas, 176.9 × 157.8 cm
National Gallery of Art, Washington, Gift
of The Mark Rothko Foundation, Inc.,
1986.43.163

143 Rotraut
Untitled, c. 1972
Mixed technique on canvas, 190 × 97 cm
Private collection

144 Rotraut
Untitled, c. 1972
Mixed technique on canvas, 116 × 89 cm
Private collection

145 John Russell
Moon globe 12-inch in diameter, 1797
Paper and brass, 51 × 45 × 47 cm
Science Museum Group, London

146 Tom Sachs
Moon Rock Box: Helpers in Need, 2008
Wood, latex paint, steel hardware, electrical
components, glass, Arten Thermohygrometer,
mixed media
20.25 × 20.5 × 7.38 cm
Private collection

147 Tom Sachs
Mars Rocks, 2012
Mars Geology Samples, ConEd Barrier,
plywood, steel, hardwae, latex paint and UV
Plexi, 187 × 186.5 × 18.5 cm
Courtesy the Artist and Sperone Westwater,
New York

148 Tom Sachs
Synthetic Mars Rocks (Sandinista), 2016
Plywood, epoxy resin, lead, latex paint and
steel, 127 × 91.5 × 23 cm
Courtesy the Artist and Sperone Westwater,
New York

149 Tom Sachs
Europa Camera System, 2018
Mixed media, 217.8 × 218.4 × 40 cm
Courtesy the Artist and Sperone Westwater,
New York

150 Tom Sachs
Europa Rock Cabinet (Sengoku), 2018
Mixed media, 120.5 × 117 × 12.5 cm
Courtesy the Artist and Sperone Westwater,
New York

151 William Safire
In Event of Moon Disaster, 18-07-1969
Paper
Richard Nixon Presidential Library and
Museum, NARA

152 Johann Hieronymus Schroeter
Selenographische Fragmente Zur Genauern
Kenntniss Der Mondfläche, 1791
Selenographic fragments towards more exact
knowledge of the lunar surface
Book, 21 × 25.5 cm
The Royal Danish Library, Copenhagen

153 David Scott
360° panorama of the Hadley-Appennine
landing site from the top hatch of the LM,
Standup EVA, Apollo 15, August 1971
Mosaic of twelve vintage gelatin silver prints
numbered in black in top margin, 26 × 140 cm
Collection Victor Martin-Malburet

154 David Scott
Portrait of James Irwin and the Rover in front
of Mount Hadley, Apollo 15, August 1971
Vintage chromogenic print on fiber-based
paper, 20.3 × 25.4 cm
Collection Victor Martin-Malburet

155 David Scott
Telephoto of Hadley Canyon's Far Wall,
Station 9A, EVA 3, Apollo 15, August 1971
Mosaic of ten vintage gelatin silver prints
numbered in black in top margin,
33 × 104 cm
Collection Victor Martin-Malburet

156 David Scott
Telephoto panorama of Hadley Rille lunar
canyon below St George Crater, Station 10,
EVA 3, Apollo 15, August 1971
Mosaic of seven vintage gelatin prints in
black numbered in top margin, 45 × 114 cm
Collection Victor Martin-Malburet

157 SEArch & CLOUDS Architecture Office
Mars Ice House, September 2015
Resin, 3D print, 33 × 45.7 × 25.4 cm
SEArch/Clouds AO

158 Alan Shepard
Fra Mauro landing site against brilliant Sun
glare, Apollo 14 February 1971
Vintage chromogenic print on resin-coated
Kodak paper, 20.3 × 25.4 cm
Collection Victor Martin-Malburet

159 Harry Shunk & Janos Kender
Yves Klein with the Blue Globe (RP 7), 1961
(org) – 2018 (exhibition print)
Photograph
Private collection

160 Kiki Smith
Moon, 1997
Lead paint on Opal Remi glass,
490 × 490 × 89 cm
The Bailey Collection, Canada

161 Kiki Smith
Tidal, 1998
Photogravure, offset lithography and
silkscreen, 50 × 320 cm
Courtesy of the Artist and Pace Gallery,
New York

162 Kiki Smith
Blue Moon III, 2011
Bronze, Moon: 124.5 × 124.5 × 6.4 cm 9 stars
Courtesy Galerie Lelong & Co.
Private collection

163 Hito Steyerl
ExtraSpaceCraft, 2016
Three channel HD video, environment
12 min, 30 sec
Courtesy the Artist and Andrew Kreps
Gallery, New York

164 August Strindberg
Celestograph I. Text on the back of the photo:
The Full Moon. Without a camera. Exposed
for 1 minute and 1½ minutes. Unfixed!,
1893-1894
Photography, 12.5 × 9 cm
Kungliga Biblioteket Sverige / National
Library of Sweden, Collection of Manuscripts

165 August Strindberg
Celestograph IV. Text on the back:
The Sun, 1893-94
Photography, 9 × 6 cm
Kungliga Biblioteket Sverige / National
Library of Sweden, Collection of Manuscripts

166 August Strindberg
Celestograph VI. Text on the back:
The Starry Sky, 1893-94
Photography, 9 × 6 cm
Kungliga Biblioteket Sverige / National
Library of Sweden, Collection of Manuscripts

167 August Strindberg
Celestograph XII, 1893-94
Photography, 9 × 6 cm
Kungliga Biblioteket Sverige / National
Library of Sweden, Collection of Manuscripts

168 August Strindberg
Celestograph XIII, 1893-94
Photography, 9 × 6 cm
Kungliga Biblioteket Sverige / National
Library of Sweden, Collection of Manuscripts

169 **August Strindberg**
Celestograph III. Text on the back: Photograph of the moon without camera or lens. The plate was in the developer for 3/4 hours and exposed to the moon that was very high in the sky. Dornach, early spring 1894, 1898-94
Photography, 12 × 8.5 cm
Kungliga Biblioteket Sverige / National Library of Sweden, Collection of Manuscripts

170 **Hiroshi Sugimoto**
Atlantic Ocean, Newfoundland, 1990
Gelatin silver print, 238.8 × 119.4 cm
Courtesy the Artist and Marian Goodman Gallery, New York

171 **Hiroshi Sugimoto**
Caribbean Sea, Yukatan, 1990
Gelatin silver print, 238.8 × 119.4 cm
Courtesy the Artist and Marian Goodman Gallery, New York

172 **Hiroshi Sugimoto**
Red Sea, Safaga, 1992
Gelatin silver print, 238.8 × 119.4 cm
Courtesy the Artist and Marian Goodman Gallery, New York

173 **Malena Szlam**
Lunar Almanac, 2013
Film, 16 mm, colour, silent 4 min, loop
LUNAR ALMANAC by Malena Szlam

174 **Unknown**
The Moon photographed by NASA's Lunar Reconnaissance Orbiter Camera: Apollo 12 landing site
NASA

175 **Unknown**
The Moon photographed by NASA's Lunar Reconnaissance Orbiter Camera: Illumination North Pole
NASA

176 **Unknown**
The Moon photographed by NASA's Lunar Reconnaissance Orbiter Camera: Craters of Different Ages
NASA

177 **Unknown**
The Moon photographed by NASA's Lunar Reconnaissance Orbiter Camera: Farside
NASA

178 **Unknown**
The Moon photographed by NASA's Lunar Reconnaissance Orbiter Camera: Nearside
NASA

179 **Unknown**
The Moon photographed by NASA's Lunar Reconnaissance Orbiter Camera: South Pole
NASA

180 **Unknown**
The limb of the Earth, Gemini 4, June 1965
Vintage chromogenic print on fiber-based Kodak paper, 20.3 × 25.4 cm
Collection Victor Martin-Malburet

181 **Unknown**
Launch from Pad 19 at Cape Kennedy, Gemini 5, 21 August 1965
Vintage chromogenic print on fiber-based Kodak paper, 20.3 × 25.4 cm
Collection Victor Martin-Malburet

182 **Unknown**
View of the giant Saturn V rocket on pad 39A at dawn, May 1966
Vintage chromogenic print on fiber-based Kodak paper, 20.3 × 25.4 cm
Collection Victor Martin-Malburet

183 **Unknown**
The first view of the Earth from the Moon, Frame 101, High Resolution, Lunar Orbiter I, 23 August 1966
Vintage gelatin silver print, 20.3 × 25.4 cm
Collection Victor Martin-Malburet

184 **Unknown**
Earthrise seen from the LM, Apollo 10, May 1969
Vintage chromogenic print on fiber-based Kodak paper, 20.3 × 25.4 cm
Collection Victor Martin-Malburet

185 **Unknown**
The CSM in lunar orbit with Michael Collins onboard, Apollo 11, July 1969
Vintage chromogenic print on fiber-based Kodak paper, 20.3 × 25.4 cm
Collection Victor Martin-Malburet

186 **Unknown**
Atlas Major, vol. 1, 88, 1708
Coloured copper engraving
The Royal Danish Library, Copenhagen

187 **Unknown**
Frederik den Femtes Atlas, bind 1, tavle 33
Frederik the Fifths Atlas, Vol. 1, Plate 33
Coloured copper engraving, 55 × c. 34 cm
The Royal Danish Library, Copenhagen

188 **Unknown**
NASA Apollo 12 LM Summary, Alternative Landing Sites, November 14, 1969, 1969
26 × 20 cm
From the private collection of Apollo 12 Astronaut Richard F. Gordon Jr.

189 **Unknown**
NASA Apollo 12 LM Lunar Surface Map, November 14, 1969, 1969
26 × 20 cm
From the private collection of Apollo 12 Astronaut Richard F. Gordon Jr.

190 **Unknown**
Lunar Sample, number 14310.221
146 gr
NASA – Lyndon B. Johnson Space Center, Houston, TX

191 **Unknown**
Moonlight mask from Ninuvak Island
Tree and feathers, diam.: 48 cm / H: 12 cm
National Museum of Denmark

192 **Unknown**
Sun mask
Tree and feathers, diam.: 51 cm, H: 13 cm
National Museum of Denmark

193 **Unknown**
Chevalier Humguffier and the Marquis de Gull making an excursion to the Moon in their new Aerial vehicle, 1784
Engraving, 17 × 19 cm
Science Museum Group, London

194 **Unknown**
Iron meteorite: Savik 1 (IIIAB), Found on Greenland in 1818
Natural History Museum of Denmark, University of Copenhagen

195 **Unknown**
Drill core from the Botanical Garden in Copenhagen
Natural History Museum of Denmark, University of Copenhagen

196 **Unknown**
Trilobite fossil
Natural History Museum of Denmark, University of Copenhagen

197 **Unknown**
Limestone – the basis of the Snowball Earth theory. Probably from Ella Island, Greenland, c. 5 × 5 cm
Natural History Museum of Denmark, University of Copenhagen

198 **Unknown**
Ketophyllum (corals) and brachiopods from Gotland
Natural History Museum of Denmark, University of Copenhagen

199 **Unknown**
Chain corals – to illustrate the Ordovician Period
20-30 cm
Natural History Museum of Denmark, University of Copenhagen

200 **Unknown**
Lunar meteorite sample
Natural History Museum of Denmark, University of Copenhagen

201 **Unknown**
Stone sample from Isua 8 (Greenland), containing evidence for early life on Earth
Natural History Museum of Denmark, University of Copenhagen

202 **Unknown**
Piece of the Allende meteorite
Natural History Museum of Denmark, University of Copenhagen

203 **Unknown**
Tetrapods from East Greenland
Natural History Museum of Denmark, University of Copenhagen

204 **Remedios Varo**
Icono, 1945
Icon
Oil, mother-of-pearl and gold leaf inlays on a wooden triptych, opened: 60 × 70 × 35 cm, closed: 60 × 39.3 × 5.5 cm
Collection MALBA, Museo de Arte Latinamericano de Buenos Aires

205 **Remedios Varo**
Portrait of Dr. Ignacio Chávez, 1957
Oil on Masonite, 94 × 61 cm
Collection of Gary and Kathie Heidenreich
Courtesy of Gallery Wendi Norris

206 **Walt Disney Studios**
Man and the Moon, 1955
Film 53 min. Footage from Man and the Moon
Courtesy of Disney Enterprises, Inc.

207 **Walt Disney Studios**
Man in Space, 1955
Film 51 min. Footage from Man in Space
Courtesy of Disney Enterprises, Inc.

208 **John Adams Whipple & George Phillips Bond**
Daguerreotype of the Moon in leather case, 1851
16.5 × 9.5 × 0.7 cm
Science Museum Group, London

209 **John Adams Whipple & George Phillips Bond**
Daguerreotype of the Moon in leather case, 1851
12.1 × 9.5 × 1 cm
Science Museum Group, London

210 **Alfred Worden**
Crescent Earth rising beyond the Moon's barren horizon, August 1971
Vintage chromogenic print on resin-coated Kodak paper, 20.3 × 25.4 cm
Collection Victor Martin-Malburet

211 **Alfred Worden**
Oblique telephoto panorama of the North Rim of Crater Pasteur on the farside of the Moon, Revolution 37, Apollo 15, August 1971
Mosaic of nine vintage gelatin silver prints numbered in black in top margin, 47 × 125 cm
Collection Victor Martin-Malburet

212 **Joseph Wright of Derby**
Virgil's Tomb, by Moonlight, 1782
Oil on canvas, 120.7 × 149.6 × 7 cm (fr)
On loan from Derby Museums Trust

213 **Joseph Wright of Derby**
Lake with Castle on a Hill, 1787
Oil on canvas, diam. 59.1 cm
Saint Louis Art Museum, Gift of Christian B. Peper

214 **Joseph Wright of Derby**
Snowdon by Moonlight, unknown
Oil on canvas, 88 × 123 cm
Courtesy of the Victoria Gallery & Museum, University of Liverpool

215 **John Young**
Ascent stage of the LM "Snoopy" flying towards the Command Module "Charlie Brown" for rendezvous, Apollo 10, May 1969
Vintage chromogenic print on fiber-based paper, 20.3 × 25.4 cm
Collection Victor Martin-Malburet

216 **Eugene Cernan, right hand glove, for extravehicular use on the lunar surface, flown, Apollo 17, 1972**
Smithsonian National Air and Space Museum, Washington

The Moon – From Inner Worlds to Outer Space

Edited by Lærke Rydal Jørgensen and Marie Laurberg
Graphic Design: Rasmus Koch Studio
Photo Editors: Sidse Buck and Kim Hansen
Translations: Glen Garner (Poul Erik Tøjner), James Manley (Marie Laurberg), Jane Rowley (Anja C. Andersen)
Proofreading: James Manley
Marie Laurberg's text has been peer-reviewed

Cover: Image from James Nasmyth's book *The Moon: Considered as a Planet, a World, and a Satellite*, London, 1874. Photo: Henni van Beek

Litho: Narayana Press
Print: Narayana Press
ISBN: 978-87-93659-08-7
Printed in Denmark 2018

www.louisiana.dk

Distributed by D.A.P. in North and Latin America and Thames & Hudson in the rest of the world

The catalogue is published on the occasion of the exhibition *The Moon – From Inner Worlds to Outer Space*
The exhibition is organized by Louisiana Museum of Modern Art, Denmark, in collaboration with Henie Onstad Kunstsenter, Norway

Louisiana Museum of Modern Art, Humlebæk
13 September 2018 – 20 January 2019
Henie Onstad Kunstsenter, Oslo
14 February – 19 May 2019

Curator: Marie Laurberg
Curatorial Assistant: Pernille Lystlund Matzen
Curatorial Coordinator / Registrar: Arne Schmidt-Petersen
Exhibition Architect: Mads Kjædegaard
Graphic Design: Marie d'Origny Lübecker
Conservator / Exhibition Producer: Børge Igor Brandt

Photo:
Marc Asekhame: p. 31 (top); Robert Bayer: p. 6; Jean-Gilles Berizzi / RMN-Grand Palais (musée du Louvre): p. 35; Ben Blackwell: p. 94; David Bordes/chateauphoto.com: p. 9, 17; Martin John Callanan: p. 5; Torben Christensen: p. 4; CloudsAO / SEArch: p. 118; Courtesy of Gallery Wendi Norris, San Francisco: p. 71 (bottom right); Jan Dixon & Emily Dixon www.WeAreTAPE.com: p. 25; Dag Fosse / KODE kunstmuseer og komponisthjem: p. 8; Fotoabteilung GNM: p. 23; Robert Gerhardt and Denis Y. Suspitsyn: p. 70 (top); Getty Images / Bettmann: p. 40; Getty Images / Field Museum Library: p. 3; Getty Images / Nick Brundle Photography: p. 42; Giuliani, Courtesy The Estate of Fabio Mauri and Hauser & Wirth: p. 88; Genevieve Hanson: p. 18; Peter Harris: p. 20; Paul Hester / Menil Collection: p. 86 (top); Roy Hewson: p. 36 (top); J.Paul Getty Trust. The Getty Research Institute, Los Angeles. (2014.R.20): p. 84; John Janca @ Artbot Photography: p. 69 (bottom); Nathan Keay / Collection Museum of Contemporary Art Chicago: p. 71 (bottom left); Pernille Klemp: p. 24; Lennart Larsen: p. 34; John Lee / Nationalmuseet: p. 47; Lobster-Fondation Groupmama Gan-Fondation Technicolor: p. 66-67; Robert R. McElroy / J. Paul Getty Trust Getty Research, Los Angeles (2014.7): p. 86 n.; John McLean: p. 36 (bottom); NASA/GSFC/Arizona State University: p. 74; Yoram Reshef Photography Studio: p. 116 (bottom); Ritzau Scanpix / Bridgeman Art Library: p. 33; Don Ross: p. 95; Christian Sardi: p. 31; Scala Archives / © 2018. Digital image, The Museum of Modern Art, New York/Scala, Florence: p. 76-77 (bottom); Peter Schibli: p. 28; Science Museum / Science & Society Picture Library: p. 80, 81 (top right); Harry Shunk and Janos Kender / J.Paul Getty Trust. The Getty Research Institute, Los Angeles. (2014.R.20): p. 104; Jakob Skou-Hansen / Statens Museum for Kunst: p. 32; Smithsonian Libraries: p. 65; Tate Images: p. 91 (top); © Trustees of the British Museum: p. 43; Henni van Beek: p. 75; J. Vogel / Max Ernst Museum: p. 69 (top); Ellen Page Wilson, courtesy Pace Gallery: p. 16; Gene Young: p. 2; Quicksilver Photographers, LLC: p. 29.

The exhibition is supported by

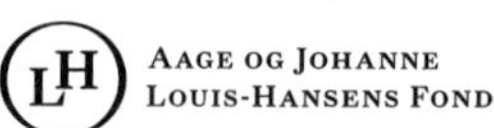

Louisiana's Main Corporate Partners

REPUBLIC OF **Fritz Hansen**®

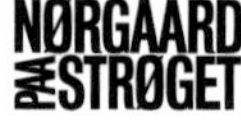

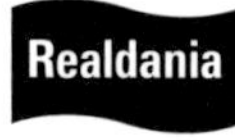

Sponsor of Louisiana's architectural exhibitions

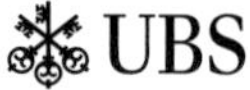

Supports programs and exhibitions at Louisiana Museum of Modern Art